Religions and Extraterrestrial Life

How Will We Deal With It?

David A. Weintraub

Religions and Extraterrestrial Life

How Will We Deal With It?

 Springer

David A. Weintraub
Department of Physics & Astronomy
Vanderbilt University
Nashville, TN, USA

SPRINGER–PRAXIS BOOKS IN POPULAR ASTRONOMY

ISBN 978-3-319-05055-3   ISBN 978-3-319-05056-0 (eBook)
DOI 10.1007/978-3-319-05056-0
Springer Cham Heidelberg New York Dordrecht London

Library of Congress Control Number: 2014941282

Printed on acid-free paper

Springer is part of Springer Science+Business Media (www.springer.com)

To
My California family, the Kennedys and Holberts:
Doris, Janelle, June, Allen, Suzy, David, Juliana, and Greg

Were there men, he was asked, living elsewhere in the universe?
"Other beings, perhaps, but not men," he answered.
Did science and religion conflict?
Not really, he said, "though it depends, of course, on your religious views."[1]

Albert Einstein
Upon his arrival in San Diego on December 31, 1930

---

[1] Isaacson, W. (2007). *Einstein*, p. 372. New York: Simon & Schuster and Brian, D. (1996). *Einstein: A Life*, p. 206. New York: John Wiley & Sons, Inc.

Acknowledgements

I offer my gratitude to Carie Lee Kennedy, Stanley Weintraub, Rodelle Weintraub, C. Robert O'Dell, and Mike Wilson, all of whom read drafts of this book and offered their wisdom and suggestions for improving the content and presentation of this book. Any errors of omission and commission that remain are all mine. I also offer my thanks to Wolf Clifton who, as a student majoring in religious studies and minoring in astronomy at Vanderbilt, spent a year conducting his own studies of the connections between extraterrestrial life and Eastern religions and whose work was valuable in guiding me toward many interesting ideas.

Contents

Part I

Discovering Extrasolar Planets

1

Once Upon a Time

> *The heavens declare the glory of God, and the firmament showeth His handiwork.*
>
> Psalm 19

ARE WE ALONE?

Once upon a time, almost two-and-a-half thousand years ago, the great Greek natural philosopher Aristotle offered nearly unassailable arguments that the Earth was the center of the celestial spheres and therefore of the entire universe. Based on the laws of physics he asserted as true, he and all of his intellectual followers knew, absolutely, that no other Earths, no other inhabitable worlds and therefore no other living, god-worshiping beings could exist anywhere in the universe but Earth. In Aristotle's universe, humanity was alone.

While other wise thinkers put forward many ideas over the next two millennia in support of the opposing view, that the universe might be full of other sentient beings living on distant worlds, Aristotle's ideas emerged intact from every intellectual skirmish except the last one. When in 1543 C. E. Copernicus hurled the Earth into orbit around the Sun, the subsequent intellectual revolution smashed Aristotle's celestial spheres to dust, rendered his laws of physics meaningless and swept the discarded remnants of the Aristotelian, geocentric universe into the trash bin of history. The other-worlds pendulum quickly swung away from the we-are-alone-in-the-universe pole and toward the we-have-lots-of-company direction.

Once upon a time, 400 years ago, the wisest scholars in the western world knew, absolutely, that the universe was swarming with inhabited worlds, other Earths. The medieval doctrine of plenitude asserted that all of God's created worlds—the Sun, the Moon,

D.A. Weintraub, *Religions and Extraterrestrial Life: How Will We Deal With It?*,
Springer Praxis Books, DOI 10.1007/978-3-319-05056-0_1,
© Springer International Publishing Switzerland 2014

the planets, the moons of Jupiter, the rings of Saturn, and all of the stars—must be inhabited. God, these scholars argued, had created these worlds and God always acts with purpose, one that we humans have the ability to understand and explain. As was self-evident to medieval Christian scholars, God's purpose in creating a universe full of worlds was to create a universe full of habitable and necessarily inhabited places. Furthermore, each and every world must be an abode for intelligent life, for creatures who could and would worship God their creator. The doctrine of plenitude even explained why stars exist: God created stars so that they might provide light and warmth for the living beings on the planets in orbit around them. The planets themselves, according to this idea which gained near universal acceptance in the seventeenth century, were created by God for the purpose of sustaining and supporting the lives of the creatures living and worshiping God in those myriad abodes.

Once upon a time, just a few decades ago, astronomer Carl Sagan, the greatest scientific popularizer in modern times of the idea that the universe must teem with life, wrote in his 1973 book *The Cosmic Connection*, "The idea of extraterrestrial life is an idea whose time has come."[1] Sagan's view of life in the universe parallels the expectations and beliefs that reverberate through modern, popular culture. Despite the absence of any scientific evidence for life beyond the Earth, in the popular imagination the universe swarms with life. Our movies, from *The War of the Worlds; The Day the Earth Stood Still; Invasion of the Body Snatchers; Alien; Star Trek*; *Star Wars*; *E.T.: The Extra-terrestrial; Close Encounters of the Third Kind; Cocoon; Independence Day; Avatar;* and even *Man of Steel* are filled with aliens. Our newspapers and magazines are checkered with reports of UFO sightings, including most famously the supposed UFO crash in Roswell, New Mexico in 1947, the lights seen in the skies above Lubbock, Texas, in 1951 and the crafts seen by pilots aboard Japan Air Lines flight 1628 in 1986. Our websites include detailed reports on alien abductions, including those of Betty and Barney Hill of New Hampshire in 1961, who claim to have been medically examined by their alien abductors, of Betty Andreasson of Massachusetts who reports that she was taken aboard alien spaceships multiple times in the 1960s, of Air Force Sergeant Charles Moody who reports that he was temporarily paralyzed by aliens in 1975 and of Australian Peter Khoury who in 1988 claims that his head was punctured by aliens with a needle. Also, our cartoons and television shows include heroes like Superman (in *Adventures of Superman*) from the planet Krypton, ALF (in *ALF*) from the planet Melmac, Mork (in *Mork & Mindy*) from the planet Ork, Spock (in *Star Trek*) from the planet Vulcan, Uncle Martin (in *My Favorite Martian*) from Mars, Commander Adama (in *Battlestar Galactica*) from the planet Caprica and the Solomon family (in *3rd Rock from the Sun*) from an unnamed planet. If books purchased, television shows watched and magazine articles read are a measure of attitude and belief, large numbers of people believe that Stonehenge was built with the help of alien engineers and that crop circles are messages from extraterrestrial visitors.

A National Geographic Channel survey of 1,114 persons conducted in 2012 indicated that 30 % of Americans (80 million people) were certain that aliens in UFOs had visited the Earth and 77 % believed some evidence exists that proves that aliens have been here. Only 48 % were not sure, and only 17 % do not believe aliens have ever visited Earth.

---

[1] Sagan, C. (1979). *The cosmic connection* (p. 198). Dell Publishing.

This study also reported that 79 % of the population thinks the government is withholding information about UFOs from the public (a result that strongly implies that 79 % of Americans think that aliens have been here, conducting business). A similar survey of 1,003 persons by the Scripps Howard News Service and Ohio University, conducted in 2008, found that one-third of adults think that intelligent aliens have visited Earth.[2] A larger, demographically broader (23,000 persons in 22 countries) Reuters/Ipsos poll taken in 2010 found that 20 % of those surveyed believe in extraterrestrials.[3] A survey of 5,886 Americans conducted in 2013 by the consumer research firm *Survata* found a similar result: 37 % of those surveyed "affirmed a belief in the existence of extraterrestrial life" while an additional 42 % were uncertain. Only 21 % denied the possibility that extraterrestrials exist. Of those responding to the survey, 55 % of atheists and 44 % of Muslims agreed with the proposition that extraterrestrial life exists. Slightly less certain were Jewish (37 %) and Hindu (36 %) respondents. Christians (32 %) were least likely to say 'yes.' Of the Christians, more than one-third of Eastern Orthodox (41 %), Roman Catholic (37 %), Methodist (37 %) and Lutheran (35 %) people surveyed were believers in extraterrestrial life, while Baptists (29 %) were below the one-third threshold.[4] Many other opinion polls taken over the last half-century reveal similar results.

Sagan wrote at a time when the theological, philosophical doctrine of plenitude had been replaced with the probabilistic, mathematical certitude of statistics. Sagan based his statistical arguments on numbers gleaned from data obtained with telescopes, in laboratory physics experiments and via space exploration. All such data taken together suggested to many scientists in the late-twentieth century and continue to suggest to a great many twenty-first century scientists that living creatures, and almost certainly intelligent creatures, must populate virtually every part of the universe. Sagan's modern, statistical, non-theistic vision posits the existence of billions of galaxies, each containing tens or hundreds of billions of stars; most of these trillions of stars likely have ten or more planets, many of which will have robust chemical mixtures as well as temperatures that are warm enough to support liquid water but not so hot as to boil away that water. Given enough time, life almost certainly will come into existence, set down roots in countless locations and evolve toward intelligence, with or without God cooking and stirring the primordial soup.

This view of the suitability of the universe for and propensity to produce life might even be right. Data derived from astrophysical theory and observations, however, can also be used to strongly suggest that we humans might be more alone than most of our forefathers and foremothers had thought. As the great physicist Enrico Fermi reputedly asked in 1950, *Where are they?*[5]

[2] More than one third of Americans believe aliens have visited Earth. *Christian Science Monitor* (2012, June 28). Retrieved from http://www.csmonitor.com/Science/2012/0628/More-than-one-third-of-Americans-believe-aliens-have-visited-Earth

[3] Do you believe in aliens? Survey says 1 in 5 believe. *Digital Journal* (2010, April 11). Retrieved from http://digitaljournal.com/article/290378

[4] survata.com/blog/science-or-sacrilege-atheists-and-agnostics-are-76-more-likely-than-christians-to-believe-in-the-existence-of-extraterrestrial-life.

[5] Jones, E. M. (1985, March). 'Where is everybody?' An account of Fermi's Question. *LA-10311-MS.*

Now, suddenly, soon after we have entered the twenty-first century, the debate about life on other worlds is about to change from the realm of thoughtful opinions to the domain of hard science. Astronomers are collecting data that will affect our understanding of and change our conversations about extraterrestrial life forever. In fact, we may soon have a very clear understanding about our place in the universe, as *we are now crossing the boundary from informed speculation to quantitatively-based knowledge as to how many Earth-like planets exist in the universe and whether any of them might harbor recognizable life forms.*

Almost every week, astronomers announce another discovery of planets around other stars. Nearly every discovery announcement is accompanied by a proclamation by a wise astronomer that this planet leads us ever closer to discovering life beyond the Earth. In 2013, a team of astronomers discovered three planets of similar size and mass to the Earth orbiting the star Gliese 667—a star located only twenty-two light years from Earth—whose orbits place them in the region around the star in which liquid water could exist on the surfaces of these planets. Astronomer Guillem Anglada-Escude commented on his team's discovery that the chance that one of these planets has conditions that can support life is "tremendous."[6] Once astronomers have identified millions of such planets where life could exist, how much more time will pass before they determine that life *does* exist?

Are we alone or do we have company out there? Whatever the answer, knowing it will trigger one of the greatest intellectual challenges in human history, not the least of which will be a theological one for terrestrial religions. Science writer Michael Chorost, writing in *The New Yorker* in 2013, asserts that "it would change everything to know that an alien civilization exists somewhere, anywhere, in the cosmos."[7] Indeed. *It would change everything.* From a religious perspective, *knowing* that humanity, that life on Earth, is but a tiny part of the cosmos could make us feel insignificant. Alternatively, as the liberal minister Parker Joss says near the end of Sagan's novel *Cosmos*, proof that we are not alone "also makes God very big."

English cosmologist and theoretical physicist Paul Davies, who is now a professor at Arizona State University, argues fairly simplistically and with attention only to western religions in his book *Are We Alone?* that our great religions will have to change dramatically, should we discover extraterrestrial beings more technologically advanced than we: "The discovery that mankind did not represent the pinnacle of evolutionary advance would prove a two-edged sword. … It is hard to see how the world's great religions could continue in anything like their present form should an alien message be received. … From the point of view of religion, it might be the case that aliens had discarded theology and religious practice long ago as primitive superstition and would rapidly convince us to do the same. … If they practiced anything remotely like a religion, we should surely soon wish to abandon our own and be converted to theirs."[8] Maybe Davies is right, and maybe he is wrong. Perhaps some religions will handle this knowledge with more flexibility and comfort than

---

[6] Landau, E. Researchers: Newly found planets might support life. Retrieved from http://www.cnn.com/2013/06/26/tech/innovation/new-habitable-planets

[7] Chorost, M. (2013, June 26). "The Seventy-Billion-Mile Telescope," in *The New Yorker* on-line.

[8] Davies, P. (1995). *Are we alone?* (pp. 54–55). New York: Basic Books.

will others. Undoubtedly, painting all religions with the same broad brush, as did Davies, is unlikely to serve as a satisfactory approach for pursuing a deep understanding of what the future may bring.

The first part of this book will introduce the discoveries in modern astronomy that suggest we are on the cusp of one of the most important intellectual revolutions in history: the discovery of life beyond the Earth. Additional details about many of these discoveries are presented in the Appendix.

I will then present an examination of the histories, scriptures and belief systems of most major religions, in order to discover whether which, if any, of these religions offer clear paths for thinking about how ready we are spiritually, theologically and scripturally for a close encounter with extraterrestrial Others.

Can we draw reasonable conclusions as to how Judaism, Roman Catholicism, Orthodox Christianity, the Church of England, mainline Protestant Christianity, fundamentalist and evangelical Christianity, Mormonism, Islam, Hinduism, Buddhism, Sikhism, Jainism, the Bahá'í Faith and a handful of other deeply-rooted religions might handle the coming intellectual revolution? Yes, we can.

For some major religions, the very concept of extraterrestrial life is built into their histories and/or scriptures; for others, ideas about extraterrestrial life are clearly irrelevant to their belief systems. For at least a few religions, evidence for the existence of extraterrestrial life would require making some adjustments, ones that can be made when the time comes, but those adjustments would not come at great theological cost; for other religious groups, important and deeply held beliefs would be fundamentally undercut by proof that extraterrestrial beings exist. In all cases, we can ask whether the beliefs and practices of each religion are human-centric and earth-centric, as this knowledge can help us understand whether a particular set of religious practices limit that religion to the Earth or if that religion could be practiced in other parts of the universe.

Might a great spiritual awakening and revolution lie just ahead of us?

2

Pluralism Through Western History

*Astronomy, like the Christian religion, if you will allow me
the comparison, has a much greater influence on our
knowledge in general, and perhaps on our manners too,
than is commonly imagined (Rittenhouse, D. (1775,
February 24). An oration delivered February 24, 1775,
before the American Philosophical Society, Philadelphia
(p. 7). Philadelphia: John Dunlap).*

David Rittenhouse, 1775 C. E.

IS THE EARTH THE ONLY LOCATION IN THE UNIVERSE
WHERE LIFE EXISTS OR CAN EXIST?

In western thought over the last several thousand years, ideas pro and con have been put forward in attempts to answer this question. Without any scientific evidence, the only strong consensus that emerged was that Aristotle's physics seemed to be right, and Aristotle's physics seemed to tie our hands in the sense that other worlds could not exist if Aristotle were right, unless of course God simply chose otherwise. This chapter offers a look at what some of our intellectual forefathers had to say on this topic, from ancient times until very recent times.

Ancient peoples likely spent very little time contemplating the existence or non-existence of other worlds, let alone whether those other worlds might be inhabited. In the west, that changed in the sixth century B. C. E. with the birth of rational science in the form of metaphysics and philosophy. In about 350 B. C. E., in his books *Physics* and *On the Heavens*, Aristotle (384–322 B. C. E.) summed up the accumulated wisdom of several centuries of Greek philosophical thought with his description of the physics at work in the universe, as understood in his time. Properly understood, the laws of physics according to Aristotle lead directly to an unequivocal answer as to whether humanity is alone in the universe: no other worlds exist and therefore humanity is alone in the universe.

D.A. Weintraub, *Religions and Extraterrestrial Life: How Will We Deal With It?*,
Springer Praxis Books, DOI 10.1007/978-3-319-05056-0_2,
© Springer International Publishing Switzerland 2014

The following logic explains how Aristotle reaches his conclusion. In an Aristotelian universe, four simple bodies—earth, water, air and fire—exist on Earth and in the Earth's atmosphere, a region that Aristotle calls the terrestrial realm. A fifth element, aether, exists only far above Earth's atmosphere, in the celestial realm. Bowls, flowers, emeralds, cows and all other real-world objects are composites made from varying proportions of the four basic building blocks of matter available in the terrestrial realm. The stars and planets, including the Sun and Moon, are made entirely of aether. Built into the very nature of each of the five elements is a purpose, and each purpose is a kind of motion.

One of the great powers of Aristotelian thinking is that Aristotle's conclusions appear to be based on observations that each and every person can and does make in his or her every-day experiences. Consequently, people can use their own senses each and every day to verify that Aristotle is right. To Aristotelians, the purpose for both earth and water is for those elements to seek their natural place at the center of the universe, that center being located at the center of the Earth. To achieve that purpose, earth and water naturally move downwards toward the center of Aristotle's universe. Indeed, in everyone's everyday experiences, objects made principally of earthy substances, like rocks and leaves and nuts, are heavy; they all fall downwards. Even if they are hurled upwards, in a direction that would be an unnatural motion for these objects, they naturally turn around and move as far downwards as they can. Likewise, fire and air seek to find their natural places as far as possible from the center of the universe, that is, as far as possible from the center of the Earth. These motions are also observable by ordinary people in their everyday experiences, as objects that one might imagine being made of fire and air are buoyant; light a fire and the sparks of fire and the hot air rise. Neither needs to be pushed upwards, as those motions are natural to those elements. Thus, both of these elements naturally move upwards, which is defined by Aristotle to be away from the center of the universe. Aether, Aristotle's fifth element, does not exist on or near the surface of the Earth. Aether only exists far above the Earth, beyond the edge of the terrestrial realm, above the sphere of fire which surrounds the sphere of air which encircles the spheres of water and earth. Beyond those four spheres that together are known as the terrestrial realm lies the realm of the aether, which fills the celestial realm. Out there, everything is made of aether, or so said Aristotle. The element aether is perfect, eternal, unchangeable and immutable; furthermore, circles have no beginnings or ends. As a result, according to Aristotle motion along the circumference of a circle is ceaseless and eternal and unchangeable and is the most natural and appropriate movement for objects made of aether. Consequently, the natural motion of anything made of aether is to move in a circle around the center of the universe (i.e., around the Earth), which is the reason objects in the heavens appear to move in circles around the Earth. Aristotle's laws of physics thus explain, naturally, why the Sun, the Moon, the planets and the stars all appear to circle the Earth.

Having established laws of physics that defined the natural, unforced motions of all objects as toward, away from or around the center of the universe, and having established from these laws of physics that the center of the Earth is identically the center of the universe, Aristotle, writing in his *On the Heavens*, pondered how earth, water, air and fire would move if another world existed. If, out there, somewhere far from the Earth, another world existed that was also made of earth and water, surely those elements would do what those elements naturally do. They would move toward the center of the universe, which is the Earth, which also means they would move upwards rather than downwards in their own worlds. In addition, particles of air and fire in this other world would move away from the Earth, away from the center of the universe and thus downwards in their own worlds. "It must be natural therefore," Aristotle wrote, "for the particles of earth in another world to move towards the centre of this one also, and for the fire in that world to move towards the circumference of this."[1]

This combination of movements, however, makes no sense, because those particles of earth in that other world also define a center of the universe, and so the universe would have two centers. Objects made of earth in our world would have to simultaneously move toward the center of the earth and toward the other center of the universe. "This is impossible," Aristotle continues, "for if it were to happen the earth would have to move upwards in its own world and the fire to the centre; and similarly earth away from the centre, as it made its way to the centre of the other, owing to the assumed situation of the worlds relatively to each other." Such a situation is self-contradictory. Something must be wrong. "Either, in fact, we must deny that the simple bodies of the several worlds have the same natures," he writes, "or if we admit it we must, as I have said, make the centre and the circumference one for all; and this means that there cannot be more worlds than one."

Since the parts of Aristotle's logic that are consistent on both the Earth and this hypothetical other world are the purposes, that is the natural motions, of the elements, the logical flaw in this scenario is that another world could exist. Therefore, Aristotle's logic demands that no other worlds, no other Earths, can exist and therefore only one habitable planet can exist in the entire universe. Such are the constraints of the laws of physics, as understood and explained by Aristotle and his followers. According to his arguments, life cannot exist anywhere other than Earth since the laws of physics preclude the existence of another abode for life anywhere else in the universe.

Although Aristotle's arguments dominated western thought from the fourth century B. C. E. until the sixteenth century C. E., other thinkers did place competing ideas into the intellectual marketplace. Epicurus (341–270 B. C. E.), who lived shortly after Aristotle and believed that the fundamental particles of the universe were atoms, argued that other worlds, full of living creatures, must exist: "Moreover, there is an infinite number of worlds, some like this world, others unlike it. For the atoms being infinite in number, as has just been proved, are borne ever further in their course. For the atoms out of which a world might arise, or by which a world might be formed, have not all been expended on one world or a finite number of worlds, whether like or unlike this one. Hence there will be nothing to hinder an infinity of worlds ... nobody can prove that in one sort of world there might not be contained ... the seeds out of which animals and plants arise and all the rest of the things we see."[2]

Unfortunately for Epicurus, his ideas about physics were less powerful than those of Aristotle and so his ideas in favor of the existence of other worlds were less convincing than Aristotle's arguments against the existence of those worlds. Nevertheless, the idea

---

[1] Aristotle. (1986). *On the heavens*, Book 1, Part VIII (p. 73) (W. K. C. Guthrie, Trans.). Cambridge: Harvard University Press.

[2] Epicurus. *Letter to Herodotus*. Retrieved from http://www.epicurus.net/en/herodotus.html

that other worlds and other-worldly creatures could exist would not go away. A few centuries later, the Roman poet Lucretius (ca. 99–55 B. C. E.), in his *De rerum natura* (*On the Nature of Things*) wrote, in agreement with Epicurus, that other worlds populated with living beings must be out there: "Of mighty things—the earth, the sea, the sky, And race of living creatures. Thus, I say, Again, again, 'tmust be confessed there are Such congregations of matter otherwhere."[3] When staking out one's intellectual position, however, it "must be confessed" is a rather weak philosophical position in comparison to a conclusion drawn from the laws of physics, i.e., from the natural philosophy of Aristotle. Half a millennium after the time of Aristotle, his ideas remained dominant.

The lack of scientific or philosophical evidence did not restrain some writers from putting anti-Aristotelian ideas into circulation, many of which were even able to offer descriptions of what extraterrestrial beings were like. Pseudo-Plutarch (the name given to the multitude of unknown authors of certain third and fourth centuries C. E. writings that were once attributed to Plutarch) described quite explicitly the characteristics of life forms on the Moon: "the moon is terraneous, is inhabited as our earth is, and contains animals of a larger size and plants of a rarer beauty than our globe affords. The animals in their virtue and energy are fifteen degrees superior to ours, emit nothing excrementitious, and the days are fifteen times longer."[4] Pseudo-Plutarch's ideas are quite consistent with the idea that creatures up there, in the heavens, are nicer or better than we are. They are prettier, more virtuous, and have better-engineered or at least more efficient bodily processes than terrestrial creatures.

From ideas such as these emerged the idea identified by historian Arthur Lovejoy as the principle of plenitude: "no genuine potentiality of being can remain unfulfilled, that the extent and abundance of the creation must be as great as the possibility of existence and commensurate with the productive capacity of a 'perfect' and inexhaustible Source, and that the world is better, the more things it contains."[5] Most of the early Christian responses to Greek and Roman ideas on the principle of plenitude were negative. The theologian Hippolytus of Rome (170–235 C. E.), the bishop Eusebius of Caesarea in Palestine (c. 260–341 C. E.), the bishop Theodoret of Cyrus (located in modern Syria) (c. 393–457 C. E.) and Saint Augustine of Hippo (located in modern Algeria) (354–430 C. E.), in agreement with Aristotle on this point, all rejected the possibility that other worlds could exist. Early Christian theologians and leaders, it would seem, preferred to believe in a universe in which we humans lived in the only place in the universe where God had chosen to create life.

Jewish scholars during the next millennium advocated for strong positions on both sides of the question. Saadia Gaon, a rabbinical scholar who lived in Babylonia in the early tenth century (882–942 C. E.), argued in favor of humankind's position of uniqueness at the center of the universe and he categorically rejected the possibility that other worlds of

---

[3] Lucretius. *On the nature of things*, *Book II* (W. E. Leonard, Trans.). Retrieved from http://classics.mit.edu/Carus/nature_things.2.ii.html

[4] Dick, S. J. (1982). *Plurality of worlds: the origins of the extraterrestrial life debate from Democritus to Kant* (p. 19). New York: Cambridge University Press.

[5] Lovejoy, A. O. (2009). *The great chain of being: A study of an idea* (p. 52). Transaction Publishers.

concentric spheres, other inhabitable worlds, could exist.[6] On the other hand, the revered Jewish scholar Moses Maimonides, a Spanish-born philosopher and physician who lived in medieval Egypt (1135–1204 C. E.), argued against the idea that humankind is all-important in the universe. Maimonides allows for the possibility that other creatures that might also be valued by God could exist elsewhere. "Consider how vast are the dimensions and how great the number of these corporeal beings," he writes in his *Guide for the Perplexed*. "The species of man is the least in comparison to the superior existents—I refer to the spheres and the stars. As far as comparison with the angels is concerned, there is in true reality no relation between man and them. Man is merely the most noble among the things that are subject to generation, namely, in this our nether world …"[7]

In Europe, for about a thousand years, Aristotle's writings were lost until after the Crusades; nevertheless, despite the fact that the young Catholic Church had some philosophical and theological problems with much of Aristotle's philosophy, as those ideas were remembered and understood, and wrapped its theology around neoPlatonic ideas, Aristotle's ideas about physics, astronomy and the universe held sway. Then, in the last centuries of the first millennium, in the eastern Mediterranean, Muslim scholars rediscovered, preserved, studied and eventually came to revere the ideas of Aristotle because, they concluded, Aristotelian philosophy was fully consistent with Islamic beliefs. When, in the twelfth century in medieval Spain, western Catholic scholars rediscovered Aristotle's ideas in Arabic translations of his works, the power of Aristotle's metaphysics and natural philosophy also became impossible for them to resist, despite the many apparent contradictions of those ideas with firmly established theological arguments. (For example, in contradiction to Genesis, where we find the words "In the beginning when God created the heavens and the earth,"[8] Aristotle's laws of physics demand that the Earth and universe are uncreated, ungenerated and eternal.) Ultimately, the thirteenth century developed as the time period during which Roman Catholic scholars built a theological structure that was largely consistent with Aristotelian concepts.

Saint Albertus Magnus (Albert the Great; 1206–1280 C. E.) fully embraced Aristotelian methods and principles and applied them to the study of Catholic scripture. By using reason in defense of faith, he merged Aristotle's philosophy of nature with the newly invented form of western scholarship known as scholasticism. Scholastic scholars used sophisticated logic to resolve apparent contradictions in authoritative texts, for example between the logical arguments put forward in an Aristotelian treatise and the theological ideas revealed in Christian scripture. This intellectual approach would dominate western scholarship for the next four centuries.

Albert the Great argued forcefully against the doctrine of double truths, in which an idea might be true in theology but a contradictory idea might be true in natural philosophy. Faith and reason, he said, could not contradict each other. If revelation offers proof that the world was created by God, yet logic offers proof that the world is eternal, then one of these

---

[6]Lamm, N. (1986). *Faith and doubt: Studies in traditional Jewish thought* (2nd ed., p. 87). New York: KTAV Publishing House.

[7]Maimonides, M. (1986). *Guide for the perplexed* as quoted in Norman Lamm, *Faith and doubt: Studies in traditional Jewish thought* (2nd ed., p. 98). New York: KTAV Publishing House.

[8]Genesis 1:1; translation from *Holy Bible New Revised Standard Version*.

ideas is wrong; if Christian truth informs believers that God could create many worlds, yet Aristotelian truth instructs scholars that only one world exists, then one of these ideas is wrong. In the early thirteenth century, Catholic believers understood that Aristotelian physics made clear that only one center of the universe—and that center is the Earth—can exist and therefore no other worlds can exist. This Aristotelian concept became a thorn in the side of thirteenth-century Catholic scholars and leaders because Aristotle appeared to be limiting God's creative power; surely, theologians argued, God could create as many worlds, populated or otherwise, as God wished. Therefore, Aristotle must be wrong. Other worlds must be possible.

How would the scholastics resolve this apparent contradiction between the authority of the Bible and the authority of Aristotle? Albert the Great's greatest student was Saint Thomas Aquinas (1225–1274 C. E.) who, in the western world, has had perhaps more intellectual influence than any thinker since Aristotle. Aquinas attempted to reconcile reason and faith, Aristotle and scripture. Aquinas used his sophisticated logic to conclude that Aristotle was right and also that scripture is inerrant: only one world exists. He asserted, however, that Aristotle's laws of physics do not limit the power of God; instead, the reverse is true: God has limited what Aristotle could observe in the real universe. God created the laws of physics. God chose to create a universe in which earth, water, fire, air and aether exist; God chose to assign to those five elements purposes defined by their natural motions; in such a universe, only one center exists. According to Aquinas, Aristotle did not create the laws of physics; he merely described for us, astutely and correctly, the laws of physics placed into the universe by God.

With this philosophical foundation, Aquinas repudiated the arguments of atomists and others who favored the existence of other worlds by claiming that those arguments had no theological basis. According to Aquinas, "those can only assert that many worlds exist who do not acknowledge any ordaining wisdom, but rather believe in chance, as Democritus, who said that this world, beside an infinite number of other worlds, was made from a casual concourse of atoms."[9]

Aquinas did not win over his theological brethren immediately. In fact, initially his ideas were widely condemned. The controversy between those who feared reason would become preeminent over faith and those who believed reason could be a dependable hand-maiden for faith rose to a climax in 1277. That year, the Bishop of Paris, Etienne Tempier, issued the Condemnation of 1277 in which he forbad faculty at the University of Paris to teach any of 219 controversial theses, including many Aristotelian arguments. Among the forbidden ideas was Article 34, "That the first cause [that is, God] could not make several worlds."[10]

The Condemnation of 1277 was effectively a trial balloon for fighting to preserve the integrity of anti-Aristotelian, Roman Catholic theology. As such, it was ineffective. After this Condemnation, as a matter of faith, at least for scholars at the University of Paris, a

---

[9]Dick, S. J. (1982). *Plurality of worlds: The origins of the extraterrestrial life debate from Democritus to Kant* (p. 27). New York, NY: Cambridge University Press.

[10]Grant, E. (2006). *Science and religion, 400 B.C. to A.D. 1500: From Aristotle to Copernicus* (p. 183). Baltimore, MD: Johns Hopkins University Press. Translated (and square brackets added) by Edward Grant.

good Catholic was compelled to believe that God had the absolute power to create any number of worlds. But by the early fourteenth century, the Condemnation of 1277 had been lifted and Thomistic theology had won the day. For the next three centuries, until the Copernican Revolution brought down the entire edifice of Aristotelian physics, Aristotle's ideas were understood as correct and consistent with Catholic theology. God had created the entire universe and chose to create but a single Earth.

LIFE EVERYWHERE: THE PRINCIPLE OF PLENITUDE IN FULL FLOWER

In practice, few scholastic writers prior to the sixteenth century believed that God chose to create more than one world, so almost no scholars believed that any other worlds existed. Agreeing with Aristotle and Aquinas was easy for most, but exceptions were the rule. Nicole Oresme (c. 1320–1382 C. E.), the Bishop of Lisieux in France and one of the most original scholastic philosophers of the fourteenth century, was one of those exceptions. He was, in general, a stern critic of Aristotle. In his *Traité du ciel et du monde* (*Treaty of Heaven and the World*), written in 1377, he argued that no argument could possibly be offered that would demonstrate, rationally, the uniqueness of the Earth. Rather than the Earth being fixed, still, at the center of the universe with the great celestial spheres turning around the Earth, Oresme asserted that the Earth could orbit the Sun.

In an Earth-centered universe, day and night occur because the Sun rises and sets. As understood by Aristotelians, the great celestial sphere that includes the Sun completes one full revolution in 24 h, thus creating day and night as the daily rotation of the celestial sphere first lifts the Sun above our eastern horizon and later drops it below our western horizon. Oresme said that no one can logically prove that this physical arrangement of the universe is correct. Instead, he argued, Aristotle's physics could be completely wrong. The Sun could be at the center of the universe and the Earth could circle the Sun (this motion of the Earth would create the annual cycle of the seasons, i.e., the year). In such a universe, the Earth would have a second motion, a daily motion, in which it would spin completely around on its axis once in 24 h, thereby making the Sun appear to rise in the morning and set in the evening: "… it is the earth that makes a daily rotation, and not the heavens. And I would like to assert the impossibility of establishing the contrary claim first by means of any observation or, secondly, by means of any rational process."[11] Without proof that Aristotle's laws of physics are true, he wrote, the Earth does not have to occupy the center of the universe and the existence of other worlds becomes possible.

Nevertheless, formulating a logical argument in which we cannot prove that other worlds do not exist is not synonymous with asserting that other worlds do exist. In the end, Oresme concluded that one cannot use logic to prove Aristotle right or to prove him wrong. Without iron-clad logical proofs in favor of or against Aristotelian physics, Oresme concluded that one must search for an answer using knowledge obtained from a source that comes from outside of humanly-derived wisdom. For Oresme, that other source of

---

[11] Danielson, D. R. (2000). *The book of the cosmos* (p. 92). Cambridge, MA: Helix Books.

knowledge is revelation, or divine wisdom, as found in Holy Scripture. Since, for Oresme, as for Aquinas and Albert the Great, only a single truth exists, divine wisdom, when interpreted correctly, is necessarily consistent with Aristotle's ideas, when understood correctly. Consequently, Oresme ultimately did not argue for the overthrow of Aristotelianism.

Nicolaus Cusanus (also, Nicholas of Cusa; 1401–1464 C. E.), one of the great German scholars of the fifteenth century who was made a cardinal in 1448, took Oresme's ideas one giant step further. Cusanus extended the idea that the existence of other worlds was possible into a claim, with absolute certainty, that such worlds do exist. Not only do other worlds exist, he asserted, they—including the Sun and the Moon—are inhabited. In agreement with Oresme, but of course with no proof, he wrote in 1440, in *De Docta ignorantia* (*On Learned Ignorance*), that the Earth moves. By implication, Aristotle's physics must be wrong. Since Aristotle's laws of physics were the basis of the argument that the Earth is the single center of the universe, Cusanus asserted that we no longer have any idea where the center of the universe is or even if the universe has a center. From that starting point, he quickly presents his ideas about the many and varied inhabitants of those countless other worlds:

> Clearly, it is actually this earth that moves, though to us it does not appear to do so ... Thus, it is as if the world system had its center everywhere and its circumference nowhere ... Since we know nothing of that whole realm, we likewise know nothing at all of its inhabitants ... Animals of one species have no concept of outsiders other than that communicated in the form of vocal expression, and this in a rather minimal way that produces nothing better than mere opinion, even after long experience. How very much less, then, can we know of the inhabitants of other worlds.
>
> Life, as it exists on earth in the form of men, animals and plants, is to be found, let us suppose, in a higher form in the solar and stellar regions. Rather than think that so many stars and parts of the heavens are uninhabited and that this earth of ours alone is peopled—and that with beings, perhaps of an inferior type—we will suppose that in every region there are inhabitants, differing in nature by rank and all owing their origin to God, who is the centre and circumference of all stellar regions... Of the inhabitants then of worlds other than our own we can know still less, having no standards by which to appraise them. It may be conjectured that in the area of the sun there exist solar beings, bright and enlightened denizens, and by nature more spiritual than such as may inhabit the moon—who are possibly lunatics—whilst those on earth are more gross and material... And we may make parallel surmise of other stellar areas that none of them lack inhabitants, as being each, like the world we live in, a particular area of one universe which contains as many such areas as there are uncountable stars.[12]

Giordano Bruno (1548–1600 C. E.), an Italian Dominican monk who on the morning of February 17, 1600 was taken to the square in Rome known as Campo de' Fiore and burned at the stake by the Roman Inquisitors for his controversial views on myriad religious and astronomical topics, was a logical, intellectual successor to Cusanus. Bruno is famous for pushing the doctrine of plenitude to its logical extreme: the universe is infinite,

---

[12] Cusanus, N. (1954). *Of learned ignorance* (G. Heron, Trans., pp. 114–115). London.

and in an infinite universe an infinite number of inhabited, Earth-like worlds must exist. In his 1584 dialogue *De l'infinito universo et Mondi* (*On the Infinite Universe and Worlds*)[13] he wrote, "the universe is infinite ... There are then innumerable suns, and an infinite number of earths revolve around those suns ... Thus there is not merely one world, one earth, one sun, but as many worlds as we see bright lights around us ..." Bruno then argues that the "fiery worlds are inhabited" as are the "watery bodies," and that even "solar creatures" exist. When one character asks "Then the other worlds are inhabited like our own?", another offers Bruno's affirming answer: "If not exactly as our own, and if not more nobly, at least no less inhabited and no less nobly."

Bruno lived and perished after the time of Polish astronomer Nicolaus Copernicus (1473–1543 C. E.) but before Copernicus's ideas had gained significant support from other scientists, let alone from religious leaders or the public. In 1543, on his deathbed, Copernicus published his book *De revolutionibus orbium coelestium* (*On the Revolutions of the Heavenly Bodies*), in which he espoused the then-radical idea that the Earth orbits the Sun, rather than vice versa. In contrast to his conceptual predecessors like Oresme and Cusanus, Copernicus did not simply offer arguments based on Aristotelian logic and scriptural exegesis; he used sophisticated mathematical and geometric tools and astronomical observations (almost none of which were his own) to model the motions of the Sun, Moon and planets.

Those models, which allow users of those tools to fairly accurately predict future positions of those celestial bodies, enabled astronomers to predict future celestial events, including lunar and solar eclipses. More importantly, by the end of the sixteenth century, Copernicus's mathematical techniques had been co-opted by astronomers in the employ of the Roman Catholic Church. Led by Christoph Clavius at the Collegio Romano in Rome, Jesuit astronomers reformed the Julian calendar, which includes a leap year every 4 years, by dropping 3 leap years out of every 400 years. The Gregorian calendar, which was first adopted by Pope Gregory XIII in 1582 and remains in use today, keeps Easter from becoming a summer holiday and prevents Christmas from becoming a spring holiday. During that same time period, however, the heliocentric hypothesis put forward by Copernicus had been rejected by almost all reputable astronomers, philosophers and theologians. Copernicus's mathematical tools were useful, the improved calendar derived therefrom was acceptable, but his big-picture cosmological idea that the Earth orbited the Sun was neither. Thus, in the year 1600, when Bruno perished, the ideas that the Earth moved and that Aristotle's laws of physics were wrong were not yet generally accepted. The stunning observations that Galileo Galilei (1564–1642 C. E.) would make with his telescope and the mathematical tour de force that Johannes Kepler (1571–1630 C. E.) would put forward with his equations for planetary motion, which together would give impetus to the nascent Copernican Revolution, were still a few years in the future.

In 1609, Galileo built his first telescope. By early 1610, he had pointed one of his telescopes at the heavens and discovered four new worlds—the objects known to us now as Io, Europa, Ganymede and Callisto. In his book *Sidereal Messenger*, published in

---

[13] Bruno, G. *On the infinite universe and worlds, 1584.* Retrieved from http://www.positiveatheism.org/hist/brunoiuw0.htm#IUWTOC

March 1610, Galileo identified them as planets "flying around the star Jupiter"[14] and named them the Medicean stars in order to honor his patron Cosimo II de' Medici, the Grand Duke of Tuscany; today, we know these four worlds as large moons that orbit the planet Jupiter.

In addition to discovering that these four newfound worlds orbit Jupiter, Galileo made several other exceedingly important new discoveries. He observed that the planet Venus, like the Moon, goes through a cycle of phases, from a tall and narrow crescent Venus to a squat but full Venus and then back again to a tall and narrow crescent Venus. When observing the Moon, he observed visual imperfections that he identified as mountains and valleys on the surface of the Moon, and when observing the Sun he noticed dark spots—sunspots— that he argued were on the surface of the Sun and that moved and changed shape and color and appeared and disappeared.

According to Aristotle's laws of physics, the Earth is the only 'center of motion.' All objects in the heavens must turn in circles centered on the Earth. Assuming that Aristotle is right, Jupiter cannot be the center of motion for the four Medicean stars, yet those four appear to circle Jupiter. Furthermore, in an Aristotelian universe, all objects in the heavens, including the Sun and Moon, are made of aether. Objects made of aether, according to Aristotle's concepts of physics, are perfect and therefore cannot experience any kind of change, including changes in shape or color. If Aristotle is right, objects made of aether cannot have visual imperfections that change. Yet the Sun has spots that move and change shape and the Moon has bright and dark spots that Galileo interpreted as the result of sun- light illuminating an object that is not in the shape of a perfectly smooth sphere. On the basis of these observations, in his 1613 book popularly known as *Letters on Sunspots,* Galileo declared Aristotle wrong and Copernicus right: "These things leave no room for doubt about the orbit of Venus. With absolute necessity we shall conclude, in agreement with the theories of the Pythagoreans and of Copernicus, that Venus revolves about the sun just as do all the other planets."[15]

When put on trial for heresy by the Roman Inquisition in 1633, Galileo would deny being a Copernican—"I do not hold this opinion of Copernicus, and I have not held it after being ordered by injunction to abandon it."[16] Despite his public denial, Galileo was, with- out a doubt, the most important scientific and philosophical catalyst for ending the 2,000 year hegemony of Aristotelian natural philosophy.

Only four decades later, Dutch astronomer Christiaan Huygens (1629–1695 C. E.) would use his own telescope to discover yet another new world, Titan, Saturn's largest moon. The telescopic proof that other worlds were centers of motion (Galileo's four newly discovered worlds orbit Jupiter, Venus orbits the Sun, Titan orbits Saturn) provided visible, robust support for the Copernican conjecture (the Moon orbits Earth while the six

[14] Galilei, G. (1989). *Sidereus nuncius* (A. Van Helden, Trans., p. 26). Chicago, IL: University of Chicago Press, title page.

[15] Galilei, G. (1957). *History and demonstrations concerning sunspots and their phenomena* (1613), as quoted in Stillman Drake's *Discoveries and opinions of Galileo* (pp. 93–94). Garden City, NY: Doubleday Anchor Books.

[16] Galilei, G. (1987). Fourth Deposition, 21 June 1633, as quoted in Maurice A. Finocchiaro. *The Galileo affair: A documentary history* (p. 287). Berkeley, CA: University of California Press.

then-known planets, including the Earth, orbit the Sun). Thus, the Copernican claim that there can be more than one center of circular motion flourished and Aristotelian physics (there is only one center of motion; that center is the Earth) were no more. Aristotle's aether, his concept that simple objects have natural motions pre-determined by purpose and his division of the universe into terrestrial and celestial realms were no more. Furthermore, with visual evidence for the existence of other worlds, the principle of plenitude leaped from Bruno's imagination into look-and-see reality.

None other than Kepler, who effectively invented mathematical astrophysics when he created the equations that describe how planets orbit the Sun in elliptical, rather than circular, orbits, was certain that Galileo's newly discovered worlds must be inhabited. According to Kepler, the mere existence of Jupiter's moons offered proof that Jupiter, itself, harbored life. In his response to *Sidereal Messenger*, Kepler wrote, "It is not improbable, I must point out, that there are inhabitants not only on the moon, but on Jupiter too."[17] Elsewhere, he wrote, "Those four little moons exist for Jupiter, not for us. Each planet in turn, together with its occupants, is served by its own satellites. From this line of reasoning, we deduce with the highest degree of probability that Jupiter is inhabited."[18] Seemingly overnight, the principle of plenitude emerged from the flames of Bruno's funeral pyre to become the standard fare for the most well-known and widely respected astronomers and physicists.

Two decades later, in 1638, the Reverend John Wilkins (1614–1672 C. E.), an Oxford graduate, the master of Trinity College, Cambridge and the Bishop of Chester, wrote (at the time anonymously) a book entitled *The Discovery of a World in the Moone, or, A Discourse Tending to Prove That 'Tis Probable There May be Another Habitable World in That Planet*. Michael J. Crowe, an historian of science, describes it as "the most influential …[o]f all the works of 'popular astronomy' in England during the seventeenth century …"[19] Wilkins, in his discourse, 'proves' his 13th proposition "that tis probable there may be inhabitants in this other World [the Moon], but of what kinde they are is uncertaine." He continues, "… the inhabitants of that world [the Moon] are not men as wee are, but some other kinde of creatures, which beare some proportion and likenesse to our natures …"[20] Wilkins leaves resolving whether these beings "are the seed of *Adam*, whether they are there in a blessed estate, or else what meanes there may be for their salvation" for others to determine.

In 1686, on the other side of the English Channel and shortly after Huygens had expanded the number of other known worlds by one, Bernard le Bovier de Fontenelle (1657–1757 C. E.) wrote his enormously influential book *Entretiens sur la pluralité des*

[17] Danielson, D. R. (2000). *The book of the cosmos* (p. 169). Cambridge, MA: Helix Books.

[18] Crowe, M J. (1999). *The extraterrestrial life debate, 1750–1900* (p. 11). Mineola, NY: Dover Publications Inc.

[19] Crowe, M J. (1999). *The extraterrestrial life debate, 1750–1900* (p. 13). Mineola, NY: Dover Publications Inc.

[20] Wilkins, J. (1638). *The discovery of a world in the Moone. Or, a discourse tending to prove that 'tis probable there may be another habitable world in that planet*. Retrieved from http://www.gutenberg.org/files/19103/19103-h/19103-h.htm

mondes (*Conversations on the Plurality of Worlds*). Fontenelle's vision of the universe includes denizens of Mercury, Saturn, Venus, Jupiter, the Moon, Jupiter's moons, comets (which he asserts are "planets which belong to a neighbouring vortex") and planets around other stars. Though he is "little acquainted" with the inhabitants of the Moon, he is confident that they are "furnished with corn, fruit, water ..." His lunarians, "being continually broiled by the excessive heat of the Sun, retire into those great caverns"; his inhabitants of Mercury "are full of fire ... are absolutely mad; ... Mercury is the bedlam of the universe"; his Venusians "resemble what I have read of the Moors of Granada, who were a little black people, scorched with the Sun, witty, full of fire, very amorous, much inclined to music and poetry ...", while his Saturnians are "very wise" and "phlegmatic: they are people who know not what it is to laugh; they take a day's time to answer the least question you can ask them; and are so very grave."[21]

Fontenelle's book was so successful that, by 1800 it had been translated into Danish, Dutch, German, Greek, Italian, Polish, Russian, Spanish and Swedish. *Entretiens* was also controversial enough in certain quarters that within a year of publication it found its way onto the *Index Librorum Prohibitorum* (The Roman Catholic Index of Prohibited Books). It was subsequently removed from the Index in 1825 and then added, once again, in 1900.[22]

Elsewhere on the continent of Europe, the German mathematician and philosopher Gottfried Wilhelm Leibniz (1646–1716 C. E.) leaned strongly toward embracing the existence of intelligent life beyond the Earth, arguing that if a form of life can possibly exist in the universe, then it must exist. The universe, he claimed, contains "an infinite number of globes, as great and greater than ours, which have as much right as it to hold rational inhabitants, though it follows not at all that they are human."[23]

The great men of ideas of the eighteenth century were at least as certain about extraterrestrial life as the great thinkers of the preceding century. In 1728 American author, scientist, inventor, postmaster and future American ambassador to France Benjamin Franklin (1706–1790 C. E.) wrote in his *Articles of Belief and Acts of Religion*, "For I believe that Man is not the most perfect Being but One, rather that as there are many Degrees of Beings his Inferiors, so there are many Degrees of Beings superior to him. ... I CONCEIVE then, that the INFINITE has created many Beings or Gods, vastly superior to Man... I conceive that each of these is exceeding wise, and good, and very powerful; and that Each has made for himself, one glorious Sun, attended with a beautiful and admirable System of Planets."[24] Sixty-two years later, Franklin wrote to Ezra Stiles, who at that time was President of Yale College, "that the most acceptable Service we can render to him [God], is doing Good to his other Children."[25] In the context of Franklin's other writings,

[21] de Fontenelle, M. (1761). *Conversations on the plurality of worlds* (Trans.). Dublin.

[22] Crowe, M. J. (1999). *The extraterrestrial life debate, 1750–1900* (p. 19). Mineola, NY: Dover Publications Inc.

[23] Crowe, M. J. (1999). *The extraterrestrial life debate, 1750–1900* (p. 28). Mineola, NY: Dover Publications Inc.

[24] Franklin, B. *Articles of belief and acts of religion*. Retrieved from http://www.historycarper.com/resources/twobf2/articles.htm

[25] Franklin, B. (1790, March 9). *Letter to Ezra stiles*. Retrieved from http://www.beliefnet.com/resourcelib/docs/44/Letter_from_Benjamin_Franklin_to_Ezra_Stiles_1.html

we can surmise that these "other Children" include intelligent beings throughout the universe.

A new tension, however, was beginning to emerge, a tension between the existence of intelligent life on other worlds and the tenets of Christianity. One of the first to strongly give voice to this tension was Philadelphian David Rittenhouse (1732–1796 C. E.). Rittenhouse, a contemporary of Franklin, was one of the most exceptional of the scientists in the American colonies and then in the young United States of America. Rittenhouse was a mathematician, clockmaker, surveyor, the first Director of the United States Mint and an astronomer, whose most famous astronomical work involved observing the transit of Venus across the face of the Sun in 1769. In his *Oration*[26] delivered to the American Philosophical Society in 1775, he insisted, like Bruno before him, "The doctrine of a plurality of worlds, is inseparable from the principles of Astronomy." He continued by pointing out that such an idea might create tension between science and Christianity: "but this doctrine is still thought, by some pious persons, and by many more I fear, who do not deserve that title, to militate against the truths asserted by the Christian religion." Rittenhouse then argued that such a conflict does not exist: "If I may be allowed to give my opinion on a matter of such importance, I must confess that I think upon a proper examination the apparent inconsistency will vanish."

While Rittenhouse did not address the many reasons why pious Christians might struggle with embracing both Christianity and extraterrestrial life, he asserted that should God want to create noble inhabitants of other worlds, God could certainly do so, even if we will never understand how this could come about: "Infinite wisdom and power, prompted by infinite goodness, may throughout the vast extent of creation and duration, have frequently interposed in a manner quite incomprehensible to us, when it became necessary to the happiness of created beings of some other rank or degree."

Rittenhouse added that astronomers have no way of knowing if "the inhabitants of the other planets may resemble man" or, now referencing the so-called Fall into sinfulness of Adam and Eve in the Garden of Eden, if "like him they were created liable to fall." He hoped that "some, if not all of them may still retain their original rectitude." He then discussed the possible governance structures of extraterrestrials: "If their inhabitants resemble man in their faculties and affections, let us suppose that they are wise enough to govern themselves according to the dictates of that reason their creator has given them, in such a manner as to consult their own and each other's true happiness on all occasions. But if on the contrary, they have found it necessary to craft artificial fabrics of government, let us not suppose they have done it with so little skill, and at such an enormous expense, as must render them a misfortune instead of a blessing. We will hope that their statesmen are patriots, and that their Kings, if that order of beings has found admittance there, have the feelings of humanity."

One of the fortunate circumstances for all extraterrestrials, including our neighbors on the Moon, according to Rittenhouse, is that we humans cannot contact or communicate *or enslave* them, as light-skinned Europeans and descendants of Europeans had done to their darker-skinned fellow humans from another continent: "Happy people! and perhaps more

---

[26] Rittenhouse, D. (1775, February 24). *An oration delivered February 24, 1775, before the American Philosophical Society* Philadelphia (pp. 19–20). Philadelphia: John Dunlap.

happy still, that all communication with us is denied. We have neither corrupted you with our vices nor injured you by violence. None of your sons and daughters, degraded from their native dignity, have been doomed to endless slavery by us in America, merely because *their* bodies may be disposed to reflect or absorb the rays of light, in a way different from *ours*. Even you, inhabitants of the Moon, situated in our very neighbourhood, are effectually secured, alike from the rapacious hand of the haughty Spaniard, and of the unfeeling British nabob."

Twenty years later and 6 months after the death of Rittenhouse, Pennsylvanian Dr. Benjamin Rush (1746–1813 C. E.), who had signed the Declaration of Independence and become a world-renowned physician, presented a eulogy for Rittenhouse. Those in attendance included members of the United States Senate and House and the Pennsylvania legislature, the luminaries in the American Philosophical Society and George Washington, the first president of the United States.[27] In his *Eulogium Intended to Perpetuate the Memory of David Rittenhouse*, Rush read nearly in its entirety the sections of Rittenhouse's *Oration* about extraterrestrial life.[28] None at the time questioned the wisdom of Rittenhouse's words or the scientific expertise embodied in those opinions, as the general belief in extraterrestrial life embraced by Rittenhouse was accepted nearly universally among the educated members of American society.

Meanwhile, on the other side of the Atlantic Ocean, the drumbeat of great European scientists and philosophers of the eighteenth century who ardently believed the solar system and universe were densely populated became a virtual who's who of famous scholars. The great French mathematician Jean le Rond d'Alembert (1717–1783 C. E.) presents himself as a pluralist, writing, "Since Saturn, Jupiter and their satellites, Mars, Venus, and Mercury are opaque bodies which receive their light from the sun, which are covered by mountains and surrounded by a changing atmosphere, it seems to follow that these *planets* have lakes, have seas … Consequently, according to many philosophers, nothing prevents us from believing that the *planets* are inhabited."[29]

Likewise, Pierre Louis Moreau de Maupertuis (1698–1759 C. E.), another French mathematician who in 1744 was the founding head of the Berlin Academy of Sciences and, briefly in the mid-1740s, director of the Académie des Sciences in Paris, presented his pluralist views, writing, "this planet [the earth] … can convince us that all the others, which appear to be of the same nature as it are not deserted globes suspended in the skies, but that they are inhabited."[30]

In 1760, Swiss physicist Leonhard Euler (1707–1783 C. E.), who was arguably the greatest mathematician of the eighteenth century, the author of over 800 papers and winner of the Grand Prize of the Paris Academy 12 times, argued similarly in favor of the existence of extraterrestrial life: "the earth, with all its inhabitants, is sometimes denominated

---

[27] Hindle, B. (1964). *David Rittenhouse* (pp. 3–4). Princeton University Press.

[28] Rush, B. (1796, December 17). *Eulogium Intended to Perpetuate the Memory of David Rittenhouse*, presented to the American Philosophical Society, Philadelphia.

[29] Ibid (p. 126)

[30] Ibid (p. 128)

a world; and every planet, nay, every one of the satellites, has an equal right to the same appellation—it being highly probable that each of these bodies is inhabited."[31]

None other than the great German philosopher Immanuel Kant (1724–1804 C. E.) wrote in 1781 in perhaps his most important work, *Critique of Pure Reason*, that he agreed with this assessment. "If it were possible to settle by any sort of experience whether there are inhabitants of at least some of the planets that we see," Kant wrote, "I might well bet everything that I have on it. Hence I say that it is not merely an opinion but a strong belief … that there are also inhabitants of other worlds."[32]

Georges Louis Leclerc, Comte de Buffon (1707–1788 C. E.; comte is the French equivalent of an English earl), the great eighteenth-century French biologist, added a new idea to the pluralist conversation. Buffon believed all planets and moons and even planetary ring systems formed hot and subsequently cooled. He further believed that each solar system locale could—and of necessity would—host life only when the temperature of that place was in a comfortable range, not too hot and not too cool. He then calculated the exact time spans and historical eras during which life could exist on each of Saturn, Jupiter, Mars, Mercury, Venus, the Moon, the moons of Saturn and Jupiter and even the rings of Saturn. According to Buffon, the entire solar system had formed exactly 74,832 years prior to the year 1775 C. E. On that basis, the Moon and Mars had once been inhabited but were not any more. All other moons and planets in the solar system, with the exception of Jupiter, were now inhabited, and Jupiter would become inhabited starting about 40,000 years from now. Therefore, according to Buffon's calculations, the solar system must be teeming with life.[33]

Buffon's approach might be the first application to identifying habitable planets of what astronomers now call the Goldilocks hypothesis. In the well-known fairy tale, Goldilocks stumbles into a seemingly empty house in the woods. On the breakfast table, she discovers three bowls of uneaten porridge that, unbeknownst to her, belong to Papa Bear, Mama Bear, and Baby Bear, all of whom are out for a stroll in the woods. Goldilocks samples the bowls of porridge and discovers that Papa Bear's porridge is too hot and Mama Bear's porridge is too cold, but Baby Bear's porridge is just right. In our own solar system today, Venus is too hot to have liquid water and thus cannot sustain life, Mars is too cold to have liquid water and thus also is unlikely to be capable of sustaining life, but the temperature of Earth (for now, at least) is just right.

The most eminent astronomers of the eighteenth century all believed that living creatures must exist beyond the Earth. William Herschel was the greatest astronomer of that century; he discovered the planet Uranus, proved binary stars exist, discovered infrared light and made the first map of the Milky Way (which he thought was a map of the entire universe). He also believed the Moon was populated with living beings—"Who can say that it is not extremely probable, nay beyond doubt, that there must be inhabitants on the

---

[31] Ibid (pp. 128–129)

[32] Kant, I. (1998). *Critique of pure reason* (p. 687). Cambridge, UK: Cambridge University Press.

[33] Crowe (1999) (p. 132)

Moon of some kind or other?"[34]—and claimed to have found visual evidence for lunar life in his observations: "I believed to perceive something which I immediately took to be *growing substances*. I will not call them Trees as from their size they can hardly come under that denomination ... My attention was chiefly directed to Mare Humorum, and this I now believe to be a forest ... I am almost convinced that those numberless small Circuses we see on the Moon are the works of the Lunarians and may be called their Towns."[35] Herschel further wrote that Mars, Saturn, Jupiter and Uranus were inhabited and that other stars most certainly had planets that "serve for the habitation of living creatures."[36]

Herschel's contemporary, the astronomer Johann Heironymous Schröter (1745–1816 C. E.), did not offer visual, telescopic evidence for extraterrestrial life, as did Herschel, but he did agree with him. Like Herschel, Schröter was a major player in the world of astronomy. He founded his own observatory in Lilienthal, Germany, with what was, at that time, the largest telescope in continental Europe and probably in the world. He then helped found the Lilienthal Detectives, an international astronomy society dedicated to finding the 'missing' planet thought to be in orbit around the Sun in between the orbits of Mars and Jupiter (during the years 1802–1807, the Lilienthal Detectives would discover three of the first four known asteroids, Pallas, Juno and Vesta, all of which do orbit the Sun in between the orbits of Mars and Jupiter). Schröter was "fully convinced that every celestial body may be so arranged physically by the Almighty as to be filled with living creatures."[37]

Schröter's colleague Johann Elert Bode (1747–1826 C. E.), another leader in the world of professional astronomers as well as one of the Lilienthal Detectives, agreed. Bode was famous for helping to formulate the Titius-Bode rule that was the *raison d'etre* for the Lilienthal Detectives. The Titius-Bode rule supposedly explained, mathematically, why the distances between the planets are what they are and predicted the existence of the 'missing' planet. He was also the long-time director of the Berlin Observatory and, for more than 50 years, the editor of what was then the most important astronomy journal in the world, the *Astronomisches Jahrbuch*. Bode firmly believed in living creatures in the Sun that he identified as Solarians, writing "Its fortunate inhabitants, say I, are illuminated by an unceasing light, the blinding brightness of which they view without injury."[38] As for the other planets, Bode asserted that "the rational inhabitants, and even the animals, plants, etc. of the other planetary bodies are characterized by forms different from those which occur on our earth."[39]

If virtually all the most eminent astronomers in the world were pluralists, then one might think that this scientific consensus represented real knowledge about the universe, but it did not. In truth, astronomers knew no more about the possibility of life beyond the Earth in the early nineteenth century than did Aristotle and his contemporaries 2,000 years

---

[34] Crowe (1999) (p. 63)
[35] Crowe (1999) (pp. 63–65)
[36] Crowe (1999) (p. 66)
[37] Crowe (1999) (p. 71)
[38] Crowe (1999) (p. 73)
[39] Crowe (1999) (p. 74)

before. In their collective ignorance, which they mistook for wisdom, the only thing that had changed was the answer.

In the mid-nineteenth century, the perceived evidence for extraterrestrials remained strong, though evidence against their existence began to appear in the work of a few serious astronomers. Friedrich Wilhelm Bessel (1784–1846 C. E.), who in 1838 made the first-ever measurement of the distance to a star beyond the Sun, using the technique known as parallax to measure the distance to 61 Cygni, was the preeminent German astronomer of his era. His observations of the Moon revealed that the transition from the bright lunar disk to the darkness of space surrounding the Moon is sharp, not gradual. He argued correctly that this result strongly suggests that the Moon has no atmosphere. He took this argument one step further by applying principles of mid-nineteenth century chemistry to his astronomical measurements. With no atmosphere, he asserted, the Moon cannot have any water: "The moon has no air; thus also no water, because without the pressure of air, water at least in the liquid state would evaporate; thus also no fire, for without air nothing can burn."[40]

The unstated implication is that without air or water, life cannot exist. Similarly, he argued that his observations indicated that the properties of other solar system bodies were markedly different from those of the Earth and that the idea that these bodies were populated was both unlikely and an idea far beyond the limits of astronomy to demonstrate. Bessel had put skepticism about extraterrestrial life back into play; in the nineteenth century, however, his position remained in the minority.

The Frenchman Camille Flammarion (1842–1925 C. E.) was the Carl Sagan of his time, the most prolific writer and popularizer of the plurality-of-inhabited-worlds hypothesis among astronomers during the nineteenth century. Flammarion was a technically strong astronomer. He observed and calculated orbits for stars in binary star systems, measured the changes in positions (known as proper motions) of a number of stars and discovered the overall motion of an entire group of stars (known as streaming) in the southern sky; he also made extensive and important studies of Mars, Venus and the Moon. Flammarion's first book, entitled *La pluralité des mondes habités* (*The Plurality of Inhabited Worlds*), appeared in 1862. *La pluralité* subsequently was revised and republished numerous times and translated into at least fifteen languages. Flammarion opined that "the forces of nature have life as their supreme end. Life is universal and eternal, for time is one of its factors. Yesterday the moon, today the earth, tomorrow Jupiter. In space there are both cradles and tombs."[41]

In an un-bylined article in Scientific American,[42] penned in 1879, Flammarion wrote that a "… telescopic examination of the planet Mars offers us the most immediate confirmation of the existence of life beyond our globe." He advised that in order to obtain the best observations of Mars, we must study Mars when it is closest to Earth and when "… the atmosphere of Mars must not be cloudy. In other words, it must be while the inhabitants of the latter planet are enjoying fine weather." As for the red color of Mars, Flammarion

---

[40] Crowe (1999) (p. 209)

[41] MacPherson, H. (1925). Camille Flammarion. *Popular Astronomy, 33*, 654.

[42] Another world inhabited like our own. *Scientific American*, May 10, 1878, p. 2787.

concludes, without equivocation, that "... the vegetation of Mars being the principal cause of this general tint ... the vegetation, whatever it is—has red for its predominant color ..."

In 1896, American astronomer Thomas Jefferson Jackson See, then working at Lowell Observatory, drew a dramatically different conclusion about life in the universe, based on his observations of double (or binary) star systems. Writing about his studies of double stars in his *Researches on the Evolution of Stellar Systems,*[43] he vastly over-interpreted his data when he concluded that the Sun and our solar system are unusual: "It is therefore impossible to determine whether the stellar systems include such bodies as the planets, and we are thus unaware of the existence of any other systems like our own. ... The solar system is rendered abnormal by the great number and small masses of its attendant bodies and by the circularity of their orbits about the large central bodies which govern their motion. The system is throughout so regular, and adjusted to such admirable conditions of stability, that among known systems it stands absolutely unique."

The author of an unattributed article in *Scientific American* written in 1905 referred to See's work and concluded that "It thus appears that so far as telescopic research has yet extended, we know of no other world suited for life outside our solar system. For some reason, our system appears to be absolutely unique in the known creation."[44] Now the pendulum was swinging back. Perhaps the solar system is a special place. Perhaps the Earth is a unique place. Yet, as before, astronomers were drawing conclusions without real data. Nothing in See's data allowed him or anyone else to draw any conclusions about the existence or non-existence of planets around other stars. Yet he did exactly that, as did those who repeated and interpreted his words.

English astronomer F. W. Henkel was correct when he wrote in *Scientific American* in 1909 that "little scientific evidence, one way or the other, exists as to such matters."[45] Speculation nevertheless was well within the bounds of reason: "Nothing seems to prevent the existence of totally different beings on every one of the planets (with the possible exception of the moon), organized and just as fitted for the conditions of their existence as we are for our own." He noted, however, that Mercury "seems at present unfitted for habitation" and that the luminous surfaces of Jupiter and Saturn argue "against the possibility of their being inhabited," though "there is no reason whatever, so far as we know, why some of their satellites, at least, should not be the abode of living beings." Closer to home, Henkel suggests that "so far as our limited knowledge extends, the evidence for the existence of living beings [on Venus], of a character not so very dissimilar from those with which we are familiar, seems as complete as we can reasonably expect." As for Mars, he was cautious, saying "Though some enthusiastic observers are convinced of the existence of rational beings, in an advanced state of civilization, inhabiting Mars, we may well pause before we arrive at this conclusion."

---

[43] See, T. J. J. (1896). *Researches on the evolution of stellar systems* (Vol. 1, p. 257). Lynn, MA: The Nichols Press.

[44] *Scientific American,* October 28, p. 334 (1905).

[45] Henkel, F. W. (1909, May 15). Life in other worlds: The interpretation of planetary markings. *Scientific American* (p. 319).

New measurements would show the wisdom in Henkel's cautionary note about Mars. Observations of Mars made in 1909 from the 14,501 foot summit of Mount Whitney in California revealed that Mars has virtually no water in its atmosphere and that the Martian polar ice caps were dominantly made of dry ice—frozen carbon dioxide. The Martian atmosphere was "a mantle of death and the shroud of animal life."[46]

Such measurements did not dissuade all astronomers. American astronomer Percival Lowell (1855–1916 C. E.) built his own observatory on Mars Hill in Flagstaff, Arizona and dedicated much of his life to the study of Mars. Based on his studies of the Martian surface, he identified north and south polar caps (he was right, Mars does have polar caps in both the north and south) as well as what he identified as canals (he was wrong, Mars does not have canals). Lowell believed the canals were enormous structures constructed by intelligent beings in order to transport water from the water-rich poles to the desert-like equatorial regions on Mars, where the water could be used for irrigation. He wrote, "It is a direct *sequitur* from this that the planet is at present the abode of intelligent, constructive life. I may say in this connection that the theory of such life upon Mars was in no way an *a priori* hypothesis on my part, but the deduced outcome of observation, and that my observations since have fully confirmed it. No other supposition is consonant with all the facts observed here."[47] Such confidence is impressive, even when the one expressing those ideas is completely wrong.

Lowell's contemporary William H. Pickering (1858–1938 C. E.) was also a member of the fading pluralist school of thought amongst late nineteenth- and early twentieth-century professional astronomers. William Pickering was a pedestrian astronomer subject to a great deal of criticism for his work, notably by his older brother Edward (1846–1919 C. E.), the more famous and very highly respected director of the Harvard College Observatory. A highlight of William Pickering's career was his 1899 discovery of Phoebe, one of the larger moons of Saturn. One of many lowlights in his observational work was his 1905 discovery of yet another Saturnian moon he named Themis; Themis does not exist. William Pickering wrote a series of papers in *Popular Astronomy* in the early 1920s, describing dark areas that, according to his measurements, moved across the lunar surface. He interpreted these as swarms of lunar insects that migrated for periods of up to 12 days; after extensive calculations, he concludes that there are about 200 million such creatures per square mile: "There is of course no reason to suppose that the lunar insects resemble locusts either in appearance or habits ... Since their environment is so different, we may say further that it is extremely improbable that they do resemble any terrestrial form of life, but since it is not worthwhile to coin a new name for them, and since they must be as we have seen of the same order of size as our insects, we will simply call them by that name, whether they have two legs, four, six, or fifty."[48]

The very highly respected Canadian-American astronomer Simon Newcomb (1835–1909 C. E.), who was a founding member and first president of the American Astronomical Society and a member of the National Academy of Sciences, used modern-sounding statistical arguments to place himself with William Pickering in the pluralist camp. In his

---

[46] The Red God of the Sky. *Scientific American Supplement*, October 23, p. 270 (1909).

[47] Lowell, P. (1907). Mars in 1907. *Nature 76*, 446.

[48] Pickering, W. H. (1924). Eratosthenes, no. 6 migration of the plats. *Popular Astronomy 32*, 392.

1905 essay "Life in the Universe," published in *The Harpers Monthly,* he points out the preponderance of astronomical and physical evidence against life existing on any of the known planets or moons in our solar system. Nevertheless, he argues that "the production of life is one of the greatest and most incessant purposes of nature." Even if "only a very small proportion of the visible worlds scattered through space are fitted to be the abode of life," the universe is so vast that the probability is high that "among hundreds of millions of such worlds a vast number are so fitted. Such being the case," he writes, "all the analogies of nature lead us to believe that, whatever the process which led to life upon this earth—whether a special act of creative power or a gradual course of development— through that same process does life begin in every part of the universe fitted to sustain it. … It is, therefore, perfectly reasonable to suppose that beings, not only animated, but endowed with reason, inhabit countless worlds in space."[49]

With the dawn of the space age and the advent of modern telescopes, the principle of plenitude lost favor in the light of evidence to the contrary. In the summer of 1950 during a visit to the Los Alamos National Laboratory, the Italian-American physicist Enrico Fermi, who was awarded the Nobel Prize in physics in 1938, famously asked, "Where are they?"[50] According to Fermi's logic, if our galaxy includes civilizations more advanced than our own, they surely would have mastered interstellar travel and would have explored the entire Milky Way galaxy. Their presence would be obvious to us. Since we do not see evidence of their presence, the absence of evidence is evidence of their absence and supports the claim that they do not exist.

By the second decade of the twenty-first century, with NASA having sent spacecraft to study the Sun, the Moon, Mercury, Venus, Mars, Jupiter, Saturn, Uranus, Neptune and Halley's Comet up close, including close-up studies of the biggest moons of Jupiter and Saturn, and with spacecraft now exploring the smaller solar system bodies Vesta, Ceres and Pluto, extraterrestrial life in our solar system has virtually no place left to hide. While the remote possibility remains that we will find primitive life forms under a Martian rock or traces of biologically generated gases in the Martian atmosphere, or that perhaps someday we will discover bacteria-like substances within a comet or in the deep subsurface water oceans of Jupiter's moon Europa, Saturn's moon Titan or Saturn's moon Enceladus, we know now that no Solarians, Lunarians, Venusians, Martians, Jupiterians, Saturnians or any other advanced life forms exist in our solar system.

If we have companions in our solar system, they are extremely primitive and are hiding very effectively. Nevertheless, most astronomers and most earthlings continue to think life is out there, if not in our solar system then somewhere further away. Our collective we-are-not-alone bias is reflected in our literature, television shows and movies, which are richly inhabited by aliens. Some of the aliens we imagine to exist are good and others are friendly, but most of them are evil or at least less moral and less ethical than we humans. None of these advanced Others, of course, are real.

---

[49] Newcomb, S. (1905, August). Life in the universe. *The Harpers Monthly,* 404.

[50] Jones, E. M. (1985). *"Where is everybody?" An account of Fermi's question,* LA-10311-MS. Retrieved from http://lib-www.lanl.gov/la-pubs/00318938.pdf

Astronomers continue their searches for evidence of radio signals from extraterrestrial beings whom we presume are more technologically advanced than we. Meanwhile, others among us continue to report sightings of UFOs, kidnappings by aliens and new crop circles that could only have been made by aliens. Many among us continue to believe that the humans who lived on our planet long before our own time could only have accomplished their great feats of engineering (e.g., building the pyramids in Egypt, erecting the monoliths at Stonehenge or raising the giant stone statues on Easter Island) with the help of supremely skilled visitors from space. Despite the fervent beliefs of many, however, there is still no broadly-accepted, verified scientific evidence that aliens exist.

In the next two chapters, we describe the dramatic discoveries of planets around other stars made by astronomers in the last two decades, of how and why the pace of those discoveries is accelerating, and of the invention of the new discipline of astrobiology. This information explains why, in the first decades of the twenty-first century, a new age is being entered, an age in which astronomers are likely to generate the data, the hard scientific evidence, that may prove once and for all that aliens either exist or do not exist in a very significant portion of our home galaxy, the Milky Way.

How, with real information in our grasp, will humans react?

3

The Discovery of the Century

We do not live in a special place in the universe.

The Cosmological Principle

PHASE ONE: DISCOVERING PLANETS

On the fifth of October, 1995, astronomers Michel Mayor and Didier Queloz of the Geneva Observatory in Switzerland made an announcement that shocked the world-wide astronomy community: "The presence of a Jupiter-mass companion to the star 51 Pegasi is inferred from observations of periodic variations in the star's radial velocity."[1]

In plain words, Mayor and Queloz had discovered that the star known as 51 Pegasi, a star in the constellation Pegasus that in all respects is nearly identical to the Sun, moves toward and away from the Earth (the *radial velocity* is the speed of the star toward and away from the Earth), over and over again, repeating this cycle of motions about every 4 days. Since stars do not bounce to and fro, the radial velocity data indicate that the star 51 Pegasi is in orbit around another object.

If we were to speak in the language of astrophysics rather than in the vernacular, we would recognize that 51 Pegasi does not orbit around an unmoving, central object. More correctly, 51 Pegasi and a second object are co-orbiting a point in space called the center-of-mass of the system. They are like two dancers circling around a spot that lies in between them on the dance floor. Similarly, in our common language, we say that the Earth orbits the Sun, but that description paints an incorrect picture in our minds of what actually occurs. The force of gravity that causes the Sun to attract the Earth also causes the Earth to attract the Sun; as a result, the Earth and Sun both orbit a point that lies somewhere in between the centers of the Earth and Sun. Because the Sun is 300,000 times more massive

---

[1] Mayor, M., & Queloz, D. (1995). A Jupiter-Mass companion to a solar-type star. *Nature, 378*, 355.

D.A. Weintraub, *Religions and Extraterrestrial Life: How Will We Deal With It?*,
Springer Praxis Books, DOI 10.1007/978-3-319-05056-0_3,
© Springer International Publishing Switzerland 2014

than the Earth, the location of the center-of-mass is 300,000 times closer to the center of the Sun than to the center of the Earth. That point is close to but is not exactly at the center of the Sun, and both the Sun (using the center of the Sun as our reference point for the motion of the Sun) and the Earth orbit this point.

In an orbiting system like that of the star 51 Pegasi, in which the mass of the unseen partner is initially unknown, astronomers can use the measurements of the light emitted by 51 Pegasi to determine its movements and to identify the location of the center-of-mass of the system; from this information they can calculate the mass of 51 Pegasi's unseen dance partner.

During the half of the orbit when the gravitational pull of the unseen partner of 51 Pegasi forces the star to come toward the Earth, the light emitted by the star travels toward the Earth at the speed of light, but the light also receives an extra push forward because of the toward-the-Earth motion of the star. This extra push gives the light just a little bit more energy than it would have if the star had not been moving when it emitted the light. Since light traveling through the near-vacuum of space travels at—no surprise here—the speed of light, this extra energy cannot speed up the light. Instead, the extra energy changes the wavelength of the light to a higher energy wavelength. When yellow light gains energy, the wavelength is changed in the direction of blue light, so such a change is called a blueshift.

During the other half of the orbit, when the unseen companion forces 51 Pegasi to travel in a direction away from the Earth, the light emitted by the star toward the Earth loses energy. Again, since the lost energy cannot affect the speed with which the light travels through space, the lost energy is manifest as a change in color of the light. If yellow light loses energy, the wavelength shifts in the direction of red light, so such a change is known as a redshift.

Mayor and Queloz had observed the light from 51 Pegasi alternately changing from redshifted to blueshifted and back again, cycling through this pattern every 4 days. They had discovered that 51 Pegasi is in a 4-day orbit with its companion. From the size of the orbit of 51 Pegasi, the speed at which it travels around its orbit, and the law of gravity, astronomers can calculate the mass of the unseen object dancing the cosmic circle dance with the star. Mayor and Queloz were able to calculate that the orbit for 51 Pegasi was too small and of too short a duration for 51 Pegasi to be in an orbit with another star. Their answer? The companion must be substellar, a tiny object in comparison to a star. The mass of the companion is about 1,000 times less than the mass of the Sun but about 300 times greater than the mass of the Earth. Such an object is much less massive than the smallest possible star but is a giant in comparison to the Earth and is, in fact, nearly identical in mass to that of the giant planet Jupiter.

On October 12, 1 week after Mayor and Queloz's announcement, astronomers Geoff Marcy (then at San Francisco State University) and Paul Butler (then at the University of California at Berkeley) announced that, using their own data obtained at Lick Observatory in California, they had confirmed Mayor and Queloz's discovery of *the first known exoplanet around a normal star*.[2] This exoplanet is now known as 51 Pegasi b. (The letter b in

---

[2] Marcy, G., & Butler, P. (1995). The planet around 51 Pegasi. *Bulletin of the American Astronomical Society, 27*, 1379.

the name indicates that this planet is the second object known in the 51 Pegasi system.) Astronomical instrumentation—the telescopes and accompanying data measurement and collection systems—had, seemingly overnight, reached the point at which dedicated planet hunters using the right observing protocols could, in only a few days or weeks, hope to detect giant planets orbiting around their host stars.

Beginning only 3 months after Mayor and Queloz made their announcement, Marcy and his team announced a series of discoveries of planets around other stars. In January 1996 they reported their discovery of a planet orbiting the star 70 Virginis[3]; in February their news was about a planet orbiting 47 Ursae Majoris[4]; in June they announced the discovery of a planet orbiting 16 Cygni B, which is one of two stars in a binary star system[5]; and in August the team let the world know about planets they had discovered orbiting each of three stars, 55 Cancri, Tau Boötis and Upsilon Andromedae.[6] In 1997, a team led by Harvard University astronomer Robert Noyes announced their discovery of a planet in orbit around Rho Coronae Borealis.[7] Astronomers collectively discovered eight more planets in 1998. Twelve new planets, including the first star with multiple planets, Upsilon Andromedae,[8] an actual planetary system discovered by Marcy's team, were discovered in 1999. By the turn of the millennium, keeping track of the new discoveries had become a full-time job.[9] Such discovery work, the first phase in astronomers' search to find and understand exoplanets, is at the cutting edge of modern astronomy and will continue for decades.

ASTROBIOLOGY

Almost immediately on the heels of the 1995 and 1996 discoveries of the first handful of planets known to orbit other stars, a second scientific announcement startled the worldwide community of astronomers and others curious about the question of how widespread life might be in the universe. The second newsworthy result was the claim announced in *Science* in August 1996 by David S. McKay and his collaborators that a meteorite found in 1984 in the Allan Hills region of Antarctica and identified as ALH 84001 had come from Mars and contained microscopic traces of evidence of ancient Martian life.[10] While the evidence that ALH 84001 originated on Mars is not controversial, the extraordinary claim that it harbors fossil evidence of life that originated on Mars has withered under a barrage

---

[3] Marcy, G., & Butler, P. (1996). A planetary companion to 70 Virginis. *Astrophysical Journal, 464*, L147.

[4] Marcy, G., & Butler, P. (1996). A planet orbiting 47 Ursae Majoris. *Astrophysical Journal, 464*, L153.

[5] Cochran, W., Hatzes, A., Butler, P., & Marcy, G. (1996). Detection of a planetary companion to 16 Cygni B. *DPS, 28*, 12.04.

[6] Butler, P., et al. (1997). Three new "51 Pegasi-type" planets. *Astrophysical Journal, 474*, L115.

[7] Noyes, R., et al. R. A planet orbiting the star rho Coronae Borealis. *Astrophysical Journal, 487*, L195.

[8] Butler, P., et al. (1999). Evidence for multiple companions to υ Andromedae. *Astrophysical Journal, 526*, 916.

[9] *exoplanets: The chronology of the discoveries*. Retrieved from http://obswww.unige.ch/~naef/RECAN/announcement.html#1995

[10] McKay, D. S., et al. (1996). Search for past life on Mars: Possible relic biogenic activity in Martian meteorite ALH84001. *Science, 273*, 924.

of strong criticisms, and most, though not all, researchers now argue that the purported biomarkers in ALH 84001 can be explained as the result of nonliving chemical processes. Nevertheless, the discovery of 51 Pegasi b combined with the claim that ALH 84001 provided proof of life whose origins lay beyond the Earth acted as a double trigger that spawned an entirely new multi-disciplinary field of study: astrobiology.

More than two decades prior to the formal birth of the field of astrobiology, meteorite researchers made some truly extraordinary discoveries that have withstood the test of time and that might provide hints about the likelihood that life exists beyond the Earth. In the late morning of September 28, 1969, residents of the small town of Murchison, in Western Australia, observed a fireball as it streaked through the sky. Fragments of this extraterrestrial impactor fell to Earth, and subsequently nearly 100 kg (220 lb) of meteoritic material in total was recovered almost immediately after it hit the ground. Because the first NASA moon landing had occurred in June 1969, barely 3 months prior to the Murchison fireball, geochemistry laboratories around the Earth were well-oiled machines, ready to study extraterrestrial materials in careful detail. Pieces of Murchison might as well have fallen right into these labs. Whereas most meteorites are known as finds, meaning that they were recovered years—sometimes millions of years—after they fell to Earth, the Murchison fragments were pristine. They were uncontaminated by the slow seepage of rainwater through the rock or by burial by sediments, both of which, over millennia, can slowly affect the mineral and chemical content of the rock; they had never suffered compression by overlying rocks or ice; they were not on the surface of the Earth long enough to have been gnawed on by animals or infiltrated by bacteria.

What did the geochemistry community learn? Murchison contains 12 % water, which is extraordinarily high for the water content of a meteorite. Murchison clearly was very fragile and had never been subjected to extreme heat or pressure before arriving on Earth—if it had been, the water would have been forced out of the rock and Murchison would have been dry. This result means the rock that became the Murchison meteorite had not been on the surface of Mars or the Moon and had never been part of a large asteroid, like Ceres or Vesta. The object that once orbited the Sun before becoming Murchison fragments was always a tiny, fragile object. Meteorites like Murchison are known as carbonaceous chondrites.

The age of Murchison, based on radioactive dating of the minerals in the meteorite, is four billion five hundred and sixty million years, the same age as other carbonaceous chondrites. That age is just a wee bit older than the ages of the Earth and the Moon (both about four billion five hundred million years old) and is thought to date the time of the birth of our entire solar system. Incredibly, inside the mineral grains that make up Murchison, geochemists working independently and on different pieces of the meteorite have found at least 74 different amino acids. Amino acids are fairly simple molecules but they are critical for life on Earth. All amino acids are composed of combinations of carbon, hydrogen and nitrogen atoms. In living materials on Earth, amino acids are the basic building blocks of proteins. The most complex amino acid (lysine) contains 25 atoms, while the simplest (alanine) contains but 13 atoms. Of the many dozens of different amino acids in Murchison, eight (including glycine, alanine, glutamic acid, valine, proline and aspartic acid) are commonly found in proteins in biological materials. Biological activity on Earth manages to incorporate 11 more of the amino acids found in Murchison. All the others are extremely rare or non-existent in the terrestrial biosphere[11] and almost certainly are extraterrestrial in origin.[12] And what about those 19? Are they terrestrial contaminants or extraterrestrial in origin? The evidence firmly points to an extraterrestrial origin. Amino acids can be 'left-handed' or 'right-handed'; the handedness refers to the internal structural shapes of the molecules and how those structural features react to light. The amino acids in Murchison are more left-handed than right-handed by a significant percentage, whereas in living terrestrial biological materials, amino acids are virtually 100 % left-handed. Once biological material is dead, the amino acids therein slowly and spontaneously revert to an equal mixture of left-handed and right-handed molecules, but that process requires hundreds of thousands of years. In addition, the ratio of carbon-12 atoms (carbon atoms containing six positively charged protons and six neutral neutrons in their nuclei) to the heavier carbon-13 atoms (carbon atoms containing six protons and seven neutrons in their nuclei) in Murchison is 88.5, whereas in all naturally occurring terrestrial materials the ratio of these two isotopes of carbon is between 90 and 92. That difference is small but scientifically significant and indicates a non-terrestrial origin for these molecules. Finally, the Murchison amino acids have more nitrogen-15 atoms (7 protons, 8 neutrons) in comparison to nitrogen-14 atoms (7 protons, 7 neutrons) than their terrestrial counterparts. All the evidence makes clear that the amino acids in the Murchison meteorite are the result of extraterrestrial chemical reactions. *They did not form on Earth.* But the evidence from the amino acids and from the carbon and nitrogen isotopic ratios isn't all that Murchison has to offer. In addition, all five of the cross-linking bases in DNA and RNA molecules—guanine, adenine, uracil, cytosine and thymine—have also been identified in Murchison and the presence of uracil of extraterrestrial origin has now been confirmed in two other meteorites (Murray and Orgueil).[13]

The Murray meteorite, which like Murchison is a carbonaceous chondrite and was also observed to fall, hit the Earth near Murray, Kentucky in 1950. About 12.6 kg of material was salvaged from the Murray site. Seventeen amino acids, balanced between right-handed and left-handed,[14] as well as several simple compounds known as polyols, which are chemical compounds that can be manufactured as sugar-free sweeteners (e.g., sorbitol) have been found in Murray fragments. The polyols cannot be biological, because they are not made biologically. But they do attest to the fact that complex, organic molecules have been synthesized naturally in space, probably in reactions between formaldehyde and water on the surfaces of small asteroids.[15]

Meteorite specialists have now identified amino acids of extraterrestrial origin in a handful of other meteorites. These discoveries lead directly to some extremely powerful

---

[11] Glavin, D. P., et al. (2010). Extraterrestrial amino acids in the Almahata Sitta meteorite. *Meteoritics & Space Science, 45,* 1695.

[12] Goldsmith, D., & Owen, T. (1993). *The search for life in the Universe* (2nd ed., p. 271). Reading, MA: Addison-Wesley Publishing.

[13] Stoks, P. G., & Schwartz, A. W. (1979). "Uracil in carbonaceous meteorites," *Nature, 282,* 709.

[14] Lawless, J. G., et al. (1971). Amino acids indigenous to the Murray Meteorite. *Science, 173,* 626.

[15] Cooper, G., et al. (2001). Carbonaceous meteorites as a source of sugar-related organic compounds for the early Earth. *Nature, 414,* 879.

questions about the origin of life on Earth. Could life on Earth have been seeded by meteorites that arrived on Earth containing complex but non-living molecules? Could meteoritic rain of amino acids with an imbalance toward left-handedness have led to the exclusive use of left-handed amino acids in life on Earth? And if life can be seeded on Earth by meteorites, could this happen elsewhere in our solar system and galaxy?

The most promising location for life elsewhere in our solar system has been thought to be the planet Mars. And though Percival Lowell's Martians never existed, and most certainly never built a global network of canals, scientists continue to search for clues that Mars, once warm and wet but now frozen, might have once had or might still have life. NASA has sent Mars Pathfinder (launched in 1996), Mars Global Surveyor (1996), Mars Climate Orbiter (1998), Mars Polar Lander (1999; this mission failed upon arrival at Mars), Mars Odyssey (2001), two Mars Exploration Rovers (Spirit and Opportunity; both in 2003), Mars Express (2003), Mars Reconnaissance Orbiter (2005), Phoenix (2007), Mars Science Laboratory (2011; also known as the Mars Rover Curiosity) and Mars Atmosphere and Volatile Evolution (2013) to the red planet to learn about the present Mars and the history of Mars and in particular to learn about the history of liquid water on the surface of Mars. Many of these missions remain active.

One possible signature in a planetary atmosphere for the evidence of life is a high level of methane gas, because sunlight should destroy methane gas in an atmosphere like that of Mars in only a few hundred years. Therefore, if any methane is present in Mars's atmosphere today, Mars must have a source that is producing methane and bubbling that gas into the atmosphere, and a plausible source for continually-generated methane is biological activity. In 2004, three different research teams reported that they had detected methane in the Martian atmosphere.[16] Michael J. Mumma, a senior scientist at NASA's Goddard Space Flight Center, who led one of the teams, suggested that microbes are likely as the origin of the methane; later studies, however, have suggested that non-living sources might be equally effective at producing the detected level of methane.[17] Furthermore, Curiosity has spent more than a year roving the surface of Mars searching for methane (among many other activities) and, as reported in the summer of 2013, has not found any, which "reduces the probability of current methanogenic microbial activity on Mars."[18] In the words of John Grotzinger, the project scientist for the Curiosity mission, "the lack of this gas 'does diminish' the possibility of methane-exhaling creatures going about their business on Mars."[19] Perhaps the 2004 reports were all in error or perhaps Mars had a biologically or geologically induced methane hiccup that year, but as of now the evidence for microbially produced methane in the atmosphere of Mars is very weak.

---

[16] Krasnopolsky, V., et al. (2004). Detection of methane in the Martian atmosphere: Evidence for life? *Icarus, 172*, 537; Formisano, V., et al. (2004). Detection of methane in the atmosphere of Mars. *Science, 306*, 1758; Mumma, M. J., et al. (2004). Detection and mapping of methane and water on Mars. *BAAS, 36*, 1127.

[17] Encrenaz, T. (2008). Search for methane on Mars: Observations, interpretation and future work. *Advances in Space Research, 42*, 1.

[18] Webster, C. R., et al. (2013). Low upper limit to methane abundance on Mars. *Science, 342*, 355.

[19] Chang, K. (2013, September 19). Life on Mars? Well, maybe not. *The New York Times*.

Astrobiology includes within its domain the search for exoplanets as well as in-depth examinations of the chemical content of meteorites and of the atmospheric content and surface geology of Mars. The field of astrobiology also encompasses down-to-Earth studies of extreme life forms known as extremophiles. Extremophiles include life forms that live deep in Earth's oceans, independent of sunlight, warmed and energized by geothermally heated water emerging from the hydrothermal vents near the mid-ocean ridges that are locations where Earth's tectonic plates are moving apart. Other extremophiles thrive at temperatures above the boiling point of water and can be found at hot springs like those in Yellowstone National Park. Yet other extremophiles grow in super acidic environments (acidophiles), super alkaline environments (alkaliphiles) or super salty environments (halophiles). One extremophile, the water bear (more precisely, the family of tardigrades), has been subjected to and has shown the ability to survive a wide variety of extreme conditions. These tiny, 1-mm-sized creatures can survive temperatures as low as −272 °C (just above absolute zero) and as high as 180 °C; they can endure a decade without water, pressures much greater than that found in the deepest ocean trenches and levels of radiation hundreds of times greater than that which would kill humans; and they have survived for 10 days in the vacuum of space. Astrobiologists have found life, in the forms of eukaryotes (single cells with nuclei) and bacteria (single cells without nuclei), in Lake Vostok, which is a very large lake (250 km in length; 30 km in width; on average 344 m in depth) buried under more than 3.5 km of Antarctic ice and rock, and which has been sealed off from Earth's atmosphere for 15 million years.[20] Environments similar to any and all of these 'extreme' environments on Earth might be found on other worlds, including on moons or other planets in our solar system. The discoveries made by astrobiologists about the breadth of environments in which life exists and persists on Earth will inform astronomers' ideas about what the chemical signatures of life on distant worlds might be.

PHASE TWO: MEASURING THE FUNDAMENTAL PROPERTIES OF EXOPLANETS

The second phase of research on exoplanets, one that has already begun, is to discern the physical and chemical properties of these planets and their host stars: What kinds of stars do exoplanets orbit (or not orbit)? Where in the galaxy are exoplanets found (or not found)? How do the sizes, masses and densities of exoplanets compare to those of planets we know well, planets like the Earth and Jupiter? Are exoplanetary systems similar to or different from our own solar system? What are the likely temperatures of exoplanets, and what fraction of them has temperatures suitable for sustaining water in the liquid phase? How fast do they spin (what are the lengths of their days)? The ultimate goal of phase-two research will be to identify the exoplanets most similar to our own planet in ways we

---

[20] Boyle, A. (2013, July 7). Antarctica's hidden Lake Vostok found to teem with life. *NBC News Report*. Retrieved from http://www.nbcnews.com/science/antarcticas-hidden-lake-vostok-found-teem-life-6C10561955; Shtarkman, Y. M., et al. (2013, July 3). Subglacial Lake Vostok (Antarctica) accretion ice contains a diverse set of sequences from aquatic, marine and sediment-inhabiting bacteria and eukarya. *PLoS One*.

believe are critical for the presence of life we might be able to identify. That is, which of the many exoplanets have sizes and masses similar to that of the Earth (or have moons with these characteristics), have atmospheres and orbit in the habitable zones (the region around a star within which liquid water can exist on the surface of a planet that has an atmosphere and a solid surface) around their stars?

Already, astronomers are taking the first steps in characterizing exoplanets. In 2011 Laurance Doyle, of the Carl Sagan Center for the Study of Life in the Universe, led a team that studied the star known as Kepler-16. They were able to determine both the size (75 % the radius of Jupiter) and mass (33 % the mass of Jupiter) of the planet known as Kepler-16b. Together, those two parameters determined the average density of the planet, which they note "is reminiscent of Saturn."[21] Also in 2011 Kaspar von Braun, of the NASA Exoplanet Science Institute at Cal Tech, and his collaborators were able to determine the radius of the super-Earth planet in orbit around the star 55 Cancri. They found that 55 Cancri e (the fifth known planet orbiting 55 Cancri) has a radius about 2.05 times greater than that of the Earth (and thus a volume about eight times greater than that of the Earth), and thus the density of the planet is in the range of 76–107 % the density of the Earth.[22] These kinds of measurements are the first steps in determining what exoplanets are made of (hydrogen and helium gases, like Saturn? rock and water like the Earth?) and thus which ones are more likely to be locations where life might have taken hold.

PHASE THREE: FINDING LIFE

After identifying the most intriguing of the known exoplanets, astronomers will attempt to determine the properties of the atmospheres of those planets. What are their temperatures? Do they have molecules in their atmospheres that would not exist in the absence of biological activity? In order to do so, we must develop observing techniques that allow us to isolate the faint light from a planet from the bright glare of its host star. Once astronomers develop the tools they need, they will be able to study in detail the light from the planets themselves.

The light beam that arrives at our telescopes from any astronomical source is a composite of the light emitted, reflected and absorbed by every kind of material in the atmosphere or on the surface of the object under study. Every kind of atom or molecule emits and absorbs light at a discrete set of wavelengths, or colors, that are a unique set of visual fingerprints for that material. The visual fingerprints for the light emitted by helium atoms, for example, are distinct from the visual fingerprints for light emitted by oxygen molecules or argon atoms. The precise combinations of colors emitted by vast numbers of atoms, molecules and chemicals have been measured in laboratories over the last century and can be compared with the spectral information obtained from exoplanets to diagnose the contents of the planets' atmospheres. Furthermore, the relative brightness of each aspect of these visual fingerprints contains information about how many atoms or

---

[21] Doyle, L. R. (2011). Kepler-16: A transiting circumbinary planet. *Science, 333*, 1602.

[22] von Braun, K., et al. (2011). 55 Cancri: Stellar astrophysical parameters, a planet in the habitable zone, and implications for the radius of a transiting super-Earth. *Astrophysical Journal, 740*, 49.

molecules of that type are emitting light and about the temperature of those atoms or molecules. Through this technique called spectroscopy, astronomers look for the distinctive signatures of different atoms and molecules in the spectra of astronomical targets and can tease out detailed information about the contents and physical properties of galaxies, stars and planets.

More than 300 years ago, Isaac Newton invented the field of spectroscopy by channeling sunlight through a glass prism. Because the glass prism refracts different colors of light at different angles as the light passes through the glass, the light exits the prism as a rainbow spectrum. In a rainbow spectrum, the incoming white light has been separated into a small handful of distinct colors—red, orange, yellow, blue, green, indigo and violet. A rainbow is a 'low resolution' spectrum. In order to identify the unique spectroscopic signatures of the materials that make up the atmospheres of stars and planets, astronomers need to separate the red part of the visual spectrum into thousands or tens of thousands or even hundreds of thousands of slightly different shades of red, and they wish to do the same with each of the other colors of the rainbow. Such a spectrum would be a 'high resolution' spectrum.

Nowadays, astronomers achieve much higher spectroscopic resolution than can be achieved with a prism by letting beams of light reflect off surfaces that include thousands of extremely tiny, closely-spaced, parallel grooves. A grooved surface (known as a grating) reflects light of different wavelengths at slightly different angles and, like a prism, spreads an incoming light beam into an outgoing light spectrum. The closer together the grooves are to each other, the more effectively they spread the light out and the more details we can see in the measured spectrum.

Using the technique of spectroscopy, we should have no problem identifying the spectroscopic signatures of the most abundant, naturally occurring gases we expect to find in the atmospheres of planets, including water, carbon dioxide, carbon monoxide, ammonia, molecular nitrogen and argon. We should also be able to identify gas species that, on Earth,[23] are present in the atmosphere at abundance levels higher than those at which they could otherwise exist without any ongoing biological activity. The most important of these biosignature gases are molecular oxygen (O_2; released during photosynthesis by plants), ozone (O_3; mostly produced when ultraviolet sunlight reacts with atmospheric molecular oxygen), methane (CH_4; produced through anaerobic respiration by organisms in the guts of termites and in the stomachs of ruminant animals, e.g., cattle, goats, deer and llamas, and in the anaerobic decay of organic matter), nitrous oxide (N_2O; emitted by denitrifying bacteria), and dimethyl sulfide ((CH_3)$_2$S; produced by bacteria and marine phytoplankton). Other molecules that could be significant biomarkers include methyl chloride (CH_3Cl; produced by algae), ammonia (NH_3; produced from the decay of organic materials), and hydrogen sulfide (H_2S; produced from the decay of organic matter). Spectra of the light we receive from exoplanets will readily reveal those planets' deepest secrets. Might they harbor any spectroscopic signatures of biological activity we would recognize?

In 2011, Jean-Michel Désert, of the Harvard-Smithsonian Center for Astrophysics, and his colleagues showed that they could learn a little bit—not very much yet, but something—about the contents of the atmosphere of the exoplanet GJ 1214b, a planet

---

[23] Woolf, N., et al. (2001). The spectrum of earthshine: A pale blue dot observed from the ground. *BAAS, 33*, 1305.

about 6 times more massive than the Earth. They did so by studying light from the parent star as that light passed through the planet's atmosphere. In particular, they measured the radius of the planet and showed that the planet's atmosphere was not dominated by hydrogen.[24]

Other research teams have begun to measure the colors and contents of the atmospheres of exoplanets. Thomas Evans, of the University of Oxford, and his team found that the planet HD 189733b has a deep blue color, which they interpret (and perhaps over-interpret) as evidence for thick clouds on the daytime side of the planet,[25] while M. Kuzuhara, of the University of Tokyo, and his team found that the planet GJ 504b appears as a dark magenta, with a relatively cool temperature of about 500 K (the boiling point of water, on the Kelvin scale, is 373 K) and a relatively cloudless atmosphere.[26] In contrast, observations of the planet Beta Pictoris b, made by M. Bonnefoy, of the Max Planck Institute for Astronomy, and his colleagues reveal that this super-Jupiter has a dusty atmosphere with a much higher temperature of about 1,700 K. These first measurements are baby steps, but before long astronomers will be making exquisitely detailed inventories of the contents and obtaining detailed temperature analyses of exoplanet atmospheres.

If the spectrum of light from even a single planet reveals the telltale spectroscopic signature of oxygen, ozone, methane, nitrous oxide or dimethyl sulfide with abundances greater than those that non-biological processes could naturally produce, if we discover even one planet with an oxygen-enhanced atmosphere, then we will have found a planet with algae or oak trees or perhaps with even more advanced life forms. Should we make such a discovery, we will know that chemical-based life, in a form we can recognize, exists somewhere other than Earth. On the other hand, if the spectra of millions of planets all offer the same answer—no biological markers in their atmospheric spectra—then we will have substantive and quantitative evidence that but for a single exception—the Earth—chemical-based life that we can recognize as alive is absent in a significant part of the Milky Way galaxy and perhaps in the entire galaxy.

THE COSMOLOGICAL PRINCIPLE

The cosmological principle, which is one of the foundational principles of modern science, encapsulates the idea that our part of the universe is not special. Instead, the piece of the universe and our galaxy that we call home is, astronomers assert, typical of the entire universe. According to the cosmological principle, Earth is a typical planet, the Sun is a normal star, our part of the galaxy is like other parts of the galaxy, and the Milky Way

---

[24] Desert, J.-M., et al. (2011). Observational evidence for a metal-rich atmosphere on the super-Earth GJ1214b. *Astrophysical Journal, 731*, L40.

[25] Evans, T. M., et al. (2013). The deep blue color of HD 189733b: Albedo measurements with Hubble space telescope/space telescope imaging spectrograph at visible wavelengths. *Astrophysical Journal, 772*, L16.

[26] Kuzuhara, M., et al. (2013). Direct imaging of a cold Jovian exoplanet in orbit around the Sun-like star GJ 504. *Astrophysical Journal, 774*, 11; *NASA News Release*. Astronomers image lowest-mass expolanet around a Sun-like star (5 Aug 2013).

galaxy and our cluster of galaxies is similar to other galaxies and clusters of galaxies. If the cosmological principle is a reasonable and rational starting point for speculating about the rest of the universe, and if the modest but significant part of the Milky Way galaxy we survey over the next century or two is void of life but for us, then we could reasonably conclude that we are alone, at least in a very significant volume of the universe. Conversely, if we find even one other planet with biological activity, then many other such life-hosting planets likely exist, not only in our part of the Milky Way but throughout our galaxy and in all other galaxies.

While astronomers do not yet have optimally-designed tools to make detailed, spectroscopic observations of the atmospheres of most exoplanets, new telescopes, measuring devices and observing techniques are under rapid development. The few exoplanets that have already been the targets of prototype spectroscopic studies are showing us both how much we can learn and how much more work we have yet to do. Soon, within two decades at the most, armies of astronomers will begin these studies in earnest. As is discussed in the next chapter, by the end of the twenty-first century, astronomers will have pointed their telescopes at millions of exoplanets and almost certainly will either have discovered firm evidence for life on a planet within a 1,000 light years of Earth or have gathered overwhelming evidence that within this substantial piece of universal real estate we are alone. Either answer will profoundly impact how we think about ourselves and about our place in the cosmos.

4

Detecting Exoplanets

Then I'll get on my knees and pray We don't get fooled again

Peter Townshend and The Who

The initial report by Mayor and Queloz that they had discovered a Jupiter-sized planet around the star 51 Pegasus was greeted with firm skepticism, even disbelief. Astronomers had driven down this road before and each time had crashed. This time, astronomers withheld their enthusiasm. We won't be fooled again, they sighed. A planet with the mass of Jupiter orbiting its star in only 4 days? Yeah, right. The closest planet to the Sun, Mercury, requires 88 days to complete a single orbit. We know that planets can't possibly orbit their stars in only 4 days! A planet so close to a star's own hot outer layers that the planet's atmosphere is heated to 1,800 °F (1,300 K)? No way. Besides, astronomers knew that planets as big as Jupiter cannot form that close to a star. Planetary systems obviously must be like our own, with the giant planets much further from their stars than the Earth is from the Sun.

Science, through the process of repeated measurements made by different teams of scientists often using different techniques, is self-correcting. Surely, in a few days or weeks or months, other astronomers would use their own data and prove that Mayor and Queloz had misunderstood their data or had failed to recognize some errors in their own methods of data analysis. This planet discovery, most astronomers were certain, would be short-lived, and Mayor and Queloz would soon be eating crow.

Astronomers, having been down this road before, knew how to steer the car and keep it on the road. In 1963, Swarthmore College astronomer Peter van de Kamp reported that he had discovered a planet with 50 % more mass than Jupiter (in the language of astronomers, 1.5 times the mass of Jupiter) in a 24-year orbit around the nearby star known as Barnard's Star.[1] Astronomers already knew that Barnard's Star was one of the closest stars

---

[1] van de Kamp, P. (1963). Astrometric study of Barnard's Star. *Astronomical Journal, 68*, 295.

D.A. Weintraub, *Religions and Extraterrestrial Life: How Will We Deal With It?*,
Springer Praxis Books, DOI 10.1007/978-3-319-05056-0_4,
© Springer International Publishing Switzerland 2014

to the Sun; because of its proximity, Barnard's Star moves in a straight line across the sky, through the years changing its position relative to other, more distant stars, while the more distant stars never move at all, appearing for all times in fixed positions relative to each other. What van de Kamp claimed to have discovered was an additional motion of Barnard's Star, a tiny wiggle in addition to the well-known straight-line movement. He made his discovery by carefully measuring the changes in position of Barnard's Star relative to other stars, using thousands of photographs of the sky taken over several decades at Swarthmore's Sproul Observatory. Van de Kamp claimed to be able to measure the apparent side-to-side wobble of Barnard's Star as the star responded to the gravitational pull of its much smaller planetary companion. In 1969, he claimed to have confirmed his own discovery and, thanks to what he believed to be the ever-improving accuracy of his data and analysis, calculated that the mass of this planet was 10 % larger than he had initially thought, some 1.7 times the mass of Jupiter.[2] Later that same year, van de Kamp reported that his observations and calculations revealed yet again that the mass of the planet needed to be revised, this time downwards, and that in fact, Barnard's Star had not one but two planets, a 1.1 Jupiter-mass planet in a 26-year orbit and a 0.8 Jupiter-mass planet in a 12-year orbit.[3]

Almost no other astronomers responded positively to van de Kamp's reported discoveries, and no others were able to replicate his results. Since the calling card of science is the reproducibility of measurements by unbiased observers, the fact that only van de Kamp could 'see' the planet around Barnard's Star was, to say the least, problematic. In 1973, van de Kamp's work was debunked twice. First, the team of George Gatewood, of the University of Pittsburgh, and Heinrich Eichhorn, of the University of South Florida, using their own observations, failed to detect any wobble in the motion of Barnard's Star.[4] Then John Hershey, a colleague of van de Kamp's at Swarthmore College, using the same photographs as used by van de Kamp, discovered that many other stars in these photographic images experienced the same periodic, side-to-side wobble as Barnard's Star.[5] Either all of these wobbling stars had identical planets or, much more likely, the data included systematic errors that van de Kamp had not recognized. The verdict? Van de Kamp's planet did not exist. The work of Gatewood, Eichhorn and Hershey effectively convinced the entire astronomy community, except for van de Kamp, that, as of 1973, the moment when we would discover the first known exoplanet lay in the future, not the past.

Nevertheless, despite being ignored by his peers van de Kamp persevered. In 1975, van de Kamp reported that his planet in the 26-year orbit was not quite as massive as he had thought in 1969—the mass had by then shrunk to only 0.4 Jupiter masses.[6] Seven years later, he reported again that the masses and orbital periods of his planets were not quite

[2] van de Kamp, P. (1969). Parallax, proper motion, acceleration, and orbital motion of Barnard's Star. *Astronomical Journal, 74*, 238.

[3] van de Kamp, P. (1969). Alternate dynamical analysis of Barnard's star. *Astronomical Journal, 74*, 757.

[4] Gatewood, G., & Eichhorn, H. (1973). An unsuccessful search for a planetary companion of Barnard's star BD +4 3561. *Astronomical Journal, 78*, 769.

[5] Hershey, J. L. (1973). Astrometric analysis of the field of AC +65 6955 from plates taken with the Sproul 24-inch refractor. *Astronomical Journal, 78*, 421.

[6] van de Kamp, P. (1975). Astrometric study of Barnard's star from plates taken with the Sproul 61-cm refractor. *Astronomical Journal, 80*, 658.

what he had previously determined: the orbital periods he now calculated were 20 and 12 years and the masses were 0.7 and 0.5 times the mass of Jupiter.[7] By this time, no other astronomers were paying attention. Jieun Choi, of the University of California at Berkeley, and his colleagues administered the *coup de grâce* to van de Kamp's planets around Barnard's Star in 2013, when they reported measurements from 25 years of observations, from 1987 through 2012, in which they find that "the habitable zone of Barnard's Star appears to be devoid of roughly Earth-mass planets or larger." Choi et al. assert that "Previous claims of planets around the star by van de Kamp are strongly refuted."[8]

Though van de Kamp's report of a planet around Barnard's Star gathered a significant amount of attention, his was not the first reported discovery of an exoplanet that did not actually exist. In fact, the year 1943 was a banner year for such reports. K. A. Strand, who, like van de Kamp, also worked at Sproul Observatory at Swarthmore College, claimed to have discovered a planet 16 times more massive than Jupiter in an eccentric orbit (when closest to its star, the planet's orbit brought it as close to the star it orbited as Venus gets to the Sun; when furthest from its star, the planet's orbit took it to a distance about 6 times further away from the central star than the Venus-Sun distance) of 4.9 years around one of the two stars in the relatively nearby binary star system 61 Cygnus.[9] Almost simultaneously, Dirk Reuyl and Erik Holmberg, working with observations obtained at the McCormick Observatory at the University of Virginia, reported that they had discovered a planet with a mass ten times greater than that of Jupiter in a 17-year orbit around one of the stars in the binary star system 70 Ophiuchi.[10]

These two near-simultaneous announcements represented an inspiring leap forward for astronomy, a technical tour de force certain to revolutionize our understanding of planets and ourselves. In an editorial in *Nature* in July of 1943, A. Hunter excitedly gushed about these two newly discovered planets, based as they were "on accurate data extending back some decades." Hunter advised readers that the claim for the planet-sized companion around 61 Cygnus is "evidently very weighty" and that the identification of a planet around 70 Ophiuchi "is even stronger."[11] Hunter suggested that after the war ended, the time would be ripe for a systematic search for other planets that would determine "within a relatively short time whether planetary companions are or are not a rare cosmic phenomenon." Hunter's prediction about the likely accomplishments of post-war astronomers was premature by half a century.

In 1957, Strand confirmed his own results, though the mass of the purported planet had shrunk to only 8 times the mass of Jupiter, the orbit was now circular and only half as large as the average size of the previously calculated orbit, making the orbit just a tad bigger than the size of Earth's orbit around the Sun, and the orbital period had shrunk just a bit,

[7] van de Kamp, P. (1982). The planetary system of Barnard's star. *Vistas in Astronomy, 26*, 141.

[8] Choi, J., et al. (2013). Precise Doppler monitoring of Barnard's Star. *Astrophysical Journal, 764*, 131.

[9] Strand, K. A. (1943). 61 Cygni as a triple system. *Publications of the Astronomical Society of the Pacific, 55*, 29.

[10] Reuyl, D., & Holmberg, E. (1943). On the existence of a third component in the system 70 Ophiuchi. *Astrophysical Journal, 97*, 41.

[11] Hunter, A. (1943). Non-solar planets. *Nature, 152*, 66.

to 4.8 years.[12] No other astronomers, however, have ever confirmed his results. Even W. D. Heintz, ironically also at Swarthmore College, reported in 1978 that "the original material was quite a weak basis for concluding the existence of an orbital effect" so small; Heintz's "much stronger data make it more likely that this result was spurious."[13] In 1995, a team led by Gordon A. H. Walker, of the University of British Columbia, determined that no planet larger than 3 Jupiter masses and with a period of less than 15 years could exist around 61 Cygnus.[14] Strand's planet quietly slipped into the dustbin of history.

The supposed planet orbiting 70 Ophiuchi discovered by Reuyl and Holmberg fared even less well. As early as 1952, none other than Strand offered a counter claim, suggesting "the present solution ... does not support the findings of Reuyl and Holmberg."[15] Almost as quickly as it had appeared, the planetary companion to 70 Ophiuchi disappeared. The only additional follow-up came from Heintz who, when dismissing Strand's planet around 61 Cygnus, also noted that his data "continued to indicate the absence of visual and photographic residuals that might support the suspected submotion repeatedly discussed 40 years ago" for 70 Ophiuchi.

First the planets around 70 Ophiuchi and 61 Cygnus, then the planet around Barnard's Star. Three exoplanets supposedly discovered. All of the discovery reports later shown to be wrong. The process of science as a self-correcting discipline had worked exceedingly well, but the astronomy community now had a well-earned reputation for ineptness in the field of exoplanet discoveries. As the old adage goes, *Fool me once, shame on you. Fool me twice, shame on me.* Having been led astray by these incorrect claims of major discoveries, the rules had changed. Extraordinary claims require extraordinary evidence, and any new claim for the discovery of the first known planet around a star other than the Sun certainly would fit into the category of an extraordinary claim; therefore, the evidence would, in future, have to be absolutely overwhelming.

In 1988, Harvard astronomer David Latham and his small team of collaborators reported their discovery of an object in orbit around the star HD 114762 that "might be a very large planet."[16] Latham's team was cautious in their published paper, suggesting that the object they had discovered, HD 114762b, "is probably a brown dwarf, and may even be a giant planet."[17] Latham was highly respected and had earned a reputation as a very careful astronomer. No one questioned the accuracy of his work. His claim passed the extraordinary evidence test, yet he made only a modest claim. The test of good science done well—reproducibility—would eventually confirm his discovery; indeed, even his original measurement that this object has a mass equal to that of 11 Jupiters or larger has been confirmed. The two words *or larger*, however, are critical for understanding the reception with which the astronomy community originally greeted Latham's discovery.

[12] Strand, K. A (1957). The orbital motion of 61 Cygni. *Astronomical Journal, 62*, 35.

[13] Heintz, W. D. (1978). Reexamination of suspected unresolved binaries. *Astrophysical Journal, 220*, 931.

[14] Walker, G. A. H., et al. (1995). A search for Jupiter-Mass companions to nearby stars. *Icarus, 116*, 359.

[15] Strand, K. A. (1952). The orbital motion of 70 Ophiuchi. *Astronomical Journal, 57*, 97.

[16] Stars' Data Hint at Other Planets (1988, August 4). *New York Times*.

[17] Latham, D., et al. (1989). The unseen companion of HD114762—A probable brown dwarf. *Nature, 339*, 38.

The measurements do not permit us to know the actual mass of the object. HD 114762b might have a mass of 11 Jupiters and thus be a giant planet, but it might also have a larger mass, perhaps 12 or 20 or 40 Jupiters, in which case it would be too massive to be considered a planet. Objects with masses greater than about 13 Jupiters are considered brown dwarfs, these being objects that are less massive than the smallest stars but more massive than the largest planets. For a short period of time after they form (tens to hundreds of millions of years, which is 'short' to astronomers), brown dwarfs are capable of generating energy and shining like stars through the process of nuclear fusion in their cores. In particular, brown dwarfs are able to fuse heavy hydrogen, known as deuterium, into helium; unlike stars, however, they cannot fuse regular hydrogen into helium. Because deuterium is about 6,000 times less abundant than regular hydrogen, brown dwarfs 'quickly' run out of fuel they can use for fusion. Once out of deuterium fuel, brown dwarfs quickly cool off and fade away. Planets, in comparison to brown dwarfs and stars, are objects that are never massive enough to permit the nuclear fusion process to take place in their cores.

Astrophysicists continue to debate the exact location of the boundary between planets and brown dwarfs. Objects in the mass range of 10–13 Jupiters are on the brown dwarf-planet border. In the case of HD 114762b, the announcement by Latham of the discovery of this brown dwarf or massive planet is example of science done well. Latham's decision to not make an extraordinary claim was well thought out. HD 114762b is an intriguing object; it might be a brown dwarf or it might be an enormous planet and the first bona fide planet ever discovered, but since it is not known (yet) and was not known in 1988 to be definitively a planet, at that time most astronomers concluded, again, that the first discovery of an exoplanet was a pot of gold yet to be found.

The year 1988 was also when Bruce Campbell, of the University of Victoria, and Gordon A. H. Walker and Stephenson Yang, of the University of British Columbia, completed a 6-year search for "substellar companions" in orbit around 16 stars similar to the Sun. They did not find what they were looking for—brown dwarfs—but they found hints for the existence of planets around 7 stars. Cautiously, they hypothesized that "companions of ~1–9 Jupiter masses are inferred … observations are continuing to confirm these [results]."[18] Four years later, Walker and Yang co-authored a paper in which they concluded that their 1988 suggestion that one of their target stars, Gamma Cephei, had a Jupiter-mass companion was wrong—instead, they said that their revised analysis "strongly implies that [the 2.5 year period] is in fact the star's period of rotation."[19] They went further in their own 1995 paper, concluding that their data showed no evidence for Jupiter-mass or larger planets "in short-period, circular orbits around some 45 nearby, solar-type stars."[20]

We know now that one of their might-be-a-planet planets, a super-Jupiter in a 2.5-year orbit around Gamma Cephei, is a bona fide planet, its existence having been confirmed in

[18] Campbell, B., Walker, G. A. H., & Yang, S. (1988). A search for substellar companions to solar-type stars. *Astrophysical Journal, 331*, 902.

[19] Walker, G. A. H., et al. (1992). γ Cephei: Rotation or planetary companion? *Astrophysical Journal, 396*, L91.

[20] Walker, G. A. H., et al. (1995). A search for Jupiter-mass companions to nearby stars. *Icarus, 116*, 359.

2003 by a team led by Artie P. Hatzes, of Thüringer Landessternwarte in Germany.[21] Another of their target stars, Beta Gemini (also known as Pollux) is now strongly suspected of having a super-Jupiter in a 590 day orbit,[22] though the evidence for this planet was clearly very weak in 1988 and 1995; a third star, Epsilon Eridani, also continues to show evidence for a Jupiter-sized planetary companion in a 7-year orbit.

The work done by the Hatzes team confirmed that the original research done by the Campbell team was work done well, but the planet orbiting Gamma Cephei was discovered in 2003, not in 1988. In 1988 Campbell and his co-authors were appropriately cautious and in 1995 Campbell and his colleagues had rejected their own evidence for the discovery of even a single exoplanet. Even though in 2012 Walker begin trying to retroactively stake a claim to having "detected the first exo-planet PRV [precise radial velocity] signature for a Jovian planet,"[23] throughout the decade of the 1990s, and even today, the world of professional astronomers quite reasonably did not embrace and has not embraced Campbell's team's 1988 might-be-a-planet claim as the discovery of first known exoplanet.

In July 1991, Matthew Bailes, Andrew Lyne and Setnam Shemar, of the University of Manchester, using data from the Jodrell Bank radio telescope in England, announced the discovery of a planet orbiting the pulsar PSR 1829–10.[24] The planet orbited in about 6 months and had a reported mass of about 10 times the mass of the Earth. This discovery was a surprise, and not a welcome one, to most astronomers interested in exoplanets.

This planet orbited a pulsar. Pulsars, which are also known as neutron stars, are dead stars, remnants of very massive stars that exploded in their old age. Each such explosion would have created a supernova, an object that would briefly have shone a billion times brighter than the Sun; most astrophysicists are convinced that the catastrophic explosion of the original star would have destroyed any planetary system that might have existed around the star before the explosion. This pulsar planet therefore probably formed after the supernova explosion, perhaps after the neutron star vaporized a companion star and created, from that process of destruction, a disk of orbiting debris in which and out of which the pulsar's planet formed.

For most astronomers, the search for planets is ultimately a quest to find life in the universe, or at least for places where life might exist. In this sense, a pulsar planet is irrelevant. It is an abnormal planet orbiting a hostile, dead star. All pulsars are sources of intense gamma rays and x-rays, which are forms of light that carry enormous amounts of energy. The intense pulsar radiation would sterilize the surface and atmosphere of any planet located close enough to a pulsar to have an orbit as short as only 6 months. Since life could not survive the harsh radiation environment of a pulsar planet, a pulsar planet is

[21] Hatzes, A. P., et al. (2003). A planetary companion to γ Cephei A. *Astrophysical Journal, 599,* 1383.

[22] Walker, G. A. H. (2012). The first high-precision radial velocity search for extra-solar planets. *New Astronomy Reviews, 56,* 9; Hatzes, A. P., et al. (2006). Confirmation of the planet hypothesis for the long-period radial velocity variations of β Geminorum. *Astronomy & Astrophysics, 457,* 335.

[23] Walker, G. A. H. (2012). The first high-precision radial velocity search for extra-solar planets. *New Astronomy Reviews, 56,* 9.

[24] Bailes, M., Lyne, A. G., & Shemar, S. L. (1991). A planet orbiting the neutron star PSR1829 - 10. *Nature, 352,* 311.

not the object for which most astronomers were looking. As real estate agents and business moguls know, location is everything, and this exoplanet is in the wrong location.

As it turned out, astronomers did not have to worry about the planet around PSR 1829–10 for long. On January 16, 1992, only 6 months after the original discovery was made public, Lyne bravely and forthrightly announced that the planet did not exist.[25] Instead, he reported that the data that appeared to indicate the presence of a planet orbiting the pulsar was actually the signature of the Earth regularly and systematically changing speed as it travels in its elliptical orbit around the Sun. The University of Manchester team had forgotten to remove this signal from their data.[26] Amazingly, Bailes, Lyne and Shemar had discovered irrefutable evidence for the Earth orbiting the Sun rather than for a planet orbiting PSR 1829–10!

Ironically, barely a week before Lyne offered his public retraction, astronomers Aleksander Wolszczan, then working at Arecibo Observatory in Puerto Rico, and Dale Frail, of the National Radio Astronomy Observatory in New Mexico, announced, to the surprise and disbelief of their colleagues, that they had discovered a planetary system, not just one planet, around a different pulsar, PSR 1257 + 12.[27] Incredibly, Wolszczan and Frail claimed to have found not just one but two, and perhaps three planets orbiting a single star, and, like the about-to-disappear non-planet around PSR 1829–10, these planets orbited a dead remnant of an exploded star. As for the planets themselves, they had Earth-like masses and Earth-like orbits, with masses of 2.8 and 3.4 times the mass of the Earth and orbits of 99 and 67 days.

Wolszczan and Frail's discovery was greeted with something less than universal applause. Not only were their planets not the first ones claimed to have been discovered in orbit around a pulsar, these planets were 2 or 3 more planets in the wrong places, in the sense that they almost certainly could not support life because of the lethal radiation from the central pulsar. In addition, not everyone was convinced that planets could even exist around pulsars, and the authors themselves were confident but not absolutely convinced of their own conclusions, writing that the data "strongly suggest" that PSR 1257 + 12 "is accompanied by a system of two more planet-sized bodies." Then the roof caved in on the non-planet around PSR 1829–10. By implication, serious doubt started to hover over the reality of the planets around PSR 1257 + 12.

Three years later, when Wolszscan reported "irrefutable evidence" for the "unambiguous detection" of both planets around PSR 1257 + 12,[28] the silence was deafening. The editors of *Science* magazine, which published Wolszscan's 1994 paper, editorialized, "The days of disappearing planets seem to be over. ... These planets aren't likely to vanish into thin air as have so many others."[29] Indeed, the planets around PSR 1257 + 12 have not

[25] Astronomer retracts his discovery of planet. *New York Times*. 16 Jan 1992.

[26] Lyne, A. G., & Bailes M. (1992). No planet orbiting PSR1829-10. *Nature, 355*, 213.

[27] Wolszczan, A., & Frail, D. A. (1992). A planetary system around the millisecond pulsar PSR1257 + 12. *Nature, 355*, 145.

[28] Wolszczan, A. (1994). Confirmation of Earth-mass planets orbiting the millisecond pulsar PSR B1257 + 12. *Science, 264*, 538.

[29] Travis, J. (1994). Pulsing star confirms more planets in the Universe. *Science, 264*, 506.

vanished, and they indeed are the first confirmed exoplanets discovered, but they also have not generated much interest because of their lack of relevance for questions related to life in the universe beyond the Earth.

As for Mayor and Queloz's planet, it also did not vanish. When Marcy and Butler announced that they had used their own data to confirm Mayor and Queloz's initial findings,[30] astronomers around the world finally celebrated. They celebrated because the planet 51 Pegasus b is a regular planet, one like Jupiter, orbiting a normal, Sun-like star, and is a planet that formed as part of the birth process of the star itself, just as the planets in our solar system formed together with the Sun. In contrast, the planets around PSR 1257 + 12 are almost certainly irrelevant for any discussions about life in the universe, and the object HD 114762b, discovered in 1988, is still not recognized as a bona fide planet, as it might be a larger object, a brown dwarf.

A BRIGHT FUTURE FOR EXOPLANET DISCOVERIES

Finding a small handful of planets around two or three stars is exciting news, but those discoveries by themselves do not mean that astronomers are on the cusp of finding life beyond the Earth. First, astronomers need to find thousands and thousands of planets that can be subjected to scrutiny. Then they need to scrutinize those planets.

This first fistful of planets did turn out to be just the proverbial tip of the iceberg. Once astronomers realized that the technique pioneered by Mayor and Queloz and by Marcy and Butler actually worked for discovering planets, dozens of other research groups jumped into the game. In addition, imaginative astronomers developed several other methods for detecting planets. Altogether, the search for exoplanets is now one of the most active and robust fields of research within astronomy. With so many projects ongoing, the number of planets discovered is already in the thousands. Having a list of exoplanets in hand that includes tens or hundreds of thousands of objects is on the foreseeable temporal horizon. Altogether, the pace of discovery in the field of exoplanet research does suggest that astronomers are, in fact, on the cusp of finding evidence regarding the existence (or non-existence) of extraterrestrial life.

The rest of this chapter provides information about what these techniques are and how they are used to detect planets. This information is the foundation for the claim made at the end of this chapter that within this century, astronomers may have scientific evidence for the existence of life in the universe beyond the Earth.

Astronomers now exploit at least six different techniques for finding planets. The most successful of these during the first decade of the exoplanet-discovery era was the radial velocity technique. Using this technique, astronomers measure the motion of a star as it responds to the gravitational tug of a planet. In the second decade of the exoplanet-discovery era, the transit technique has surpassed the radial-velocity technique in its proficiency for discovering exoplanets. With this technique, astronomers measure the drop in brightness of

---

[30] Wilford, J. N. (1995, October 20). 2 sightings of planet orbiting a Sun-like star challenge notions that Earth is unique. *New York Times*; Marcy, G. W., & Butler, R. P. (1995). The planet around 51 Pegasi. *Bulletin of the American Astronomical Society, 27*, 1379.

a star when a planet passes in front of it. Three other techniques—gravitational microlensing, direct imaging and timing—so far have each contributed more than a dozen planets to our collection of known exoplanets and a sixth approach, astrometry, has been used to possibly identify one planet.

THE RADIAL VELOCITY TECHNIQUE

In a normal conversation, we might say, "The Earth orbits the Sun." With these simple words, we mislead ourselves. The Earth appears to orbit a star whose location is fixed in space. The Sun, however, does not sit still in space. Physics, specifically the law of gravity, tells another story: the Sun moves, but the motion of the Sun is 300,000 times smaller than the motion of the Earth because the mass of the Sun is 300,000 times greater than the mass of the Earth.

Let's imagine that, instead of the Earth, the object orbiting the Sun at a distance of 150 million kilometers (93 million miles) was another star, identical in every way to the Sun. Let's call this other star Spud. Why would Spud orbit the Sun? Shouldn't the Sun orbit Spud? If the less massive object (e.g., the Earth) orbits the more massive object (e.g., the Sun), what happens when the two objects (the Sun and Spud) are identical in mass? The answer is as simple as gravity. Neither object orbits the other. Instead, both the Sun and Spud would orbit the *center of mass* of the system, which would be located exactly half way between them.

The Earth and Sun also both orbit the center of mass of the Earth-Sun system, but that center of mass is located 300,000 times closer to the center of the Sun than to the center of the Earth. If we divide the distance between the center of the Earth and the center of the Sun by 300,000, we find that the center of mass of the Earth-Sun system is located deep inside the Sun, only 500 km from the center of the Sun. Because the radius of the Sun is 696,350 km,[31] this center-of-mass location is 695,850 km below the surface of the Sun; thus, the Sun doesn't appear to move much. Because the Sun barely moves, an observer who doesn't look or measure carefully enough would assume that the Sun is fixed in place while the Earth orbits the stationary Sun.

If we were able to watch the Earth-Sun system very carefully from a vantage point located many light-years distant from the Earth-Sun system, we would see the Sun navigate a 1-year orbit with an orbital diameter of only 1,000 km and circumference of just over 3,000 km. The Sun would move slowly, at a speed of only 8.6 km/day, equivalent to 360 m/h or 10 cm/s (about 4 in/s). The Earth, in comparison, would dash around its much bigger orbit (circumference 940 million kilometers) at the astonishing speed of about 30 km/s (almost 19 miles/s). From our imagined distant vantage point looking back at the Earth-Sun system, we would notice that during one moment as the Sun traveled around its tiny annual orbit, it would move *toward us* at a top speed of 10 cm/s. Then, over a time span of 3 months, its motion toward us would appear to slow down. The Sun, of course,

[31]Emilio, M., et al. (2012). Measuring the solar radius from space during the 2003 and 2006 Mercury transits. *Astrophysical Journal, 750,* 135.

would not be slowing down. Instead, it would be changing direction as it follows its circular orbit, slowly turning until it was moving across our line of sight (let's say from left to right) rather than directly toward us. Were we able to conduct these observations of the Sun from our distant observing location, astronomers could easily measure the component of the Sun's velocity that would be directly toward or away from our imaginary observing position. The toward/away from us velocity is known as the *radial velocity*, and by this measure the Sun would appear to be decelerating from 10 cm/s to 0 cm/s. (The sideways motion of a star, known as the star's proper motion, is very difficult for astronomers to measure and can only be measured for very nearby stars, like Barnard's Star, which are moving across our line of sight at very high angular speeds.)

Once the Sun reached an apparent speed (toward us) of 0 cm/s, and as it continued to travel in its circular path, it would begin to turn away from us. Its sideways motion would decrease while its motion away from us would gradually increase. Three months after we noticed the Sun's radial velocity dropping all the way to 0 cm/s, it would be moving away from us at a top speed of 10 cm/s. Beginning at this moment and continuing over the next 3 months, the Sun would follow its circular path until it was moving strictly sideways again (this time from right to left), as seen by us. Finally, it would turn again such that it was moving toward us at a speed of 10 cm/s. At this point in its orbit, it would have completed one full orbit in 12 months. Because we are only able to measure the Sun's speed toward and away from us, we would find that that Sun's radial velocity would start at −10 cm/s (the minus sign indicates that the velocity of the Sun is toward us) and then would gradually increase to zero. Next, the velocity would increase all the way to +10 cm/s (the plus sign indicates that the velocity of the Sun is away from us), at which time it would gradually decrease again to zero and then decrease all the way to −10 cm/s.

Radial velocities can be measured through the phenomenon known as the Doppler shift. Highway patrol officers use the Doppler shift to measure the speeds of cars on the interstate; baseball scouts take advantage of the Doppler shift to determine whether a left-handed, flame-throwing, high school phenom pitches a fastball at a top speed of 91 or 97 miles/h; and astronomers use the Doppler shift to measure the motions of astrophysical objects toward or away from us. Measurements of motions as small as about 1 m/s (100 cm/s) are now just at the limit of astronomers' technical skills, but while motions as small as 10 cm/s are currently just beyond the reach of practicing astronomers, within a few years astronomers will have developed new tools that will enable them to reach that limit. At that time in the very near future, astronomers will be able to use the radial velocity technique to detect Earth-like planets in 1-year orbits around Sun-like stars.

Each pair of celestial dancers, star and planet, do-si-do their partners as they use their masses to exert gravitational pulls on each other. The strengths of their mutual pulls help determine the velocities at which they orbit each other. Knowing this, astronomers can turn this information around: if we know the velocity at which the star is orbiting around its planetary partner, and if we know the mass of the star, we can use the law of gravity to calculate the mass of the planet. Thus, the radial velocity method allows astronomers to directly measure the masses of the planets they detect.

What are the current limits of the radial velocity technique? Could astronomers looking back at the Sun detect the presence of Jupiter in orbit around the Sun? The mass of Jupiter is 318 times greater than the mass of the Earth. If Jupiter (instead of the Earth) orbited the Sun at a distance of 150 million kilometers, Jupiter would yank on the Sun and cause the Sun to move toward and away from our distant observers at a speed 318 times greater than the speed at which the Sun moves when tugged on by the Earth. Instead of moving at a maximum speed toward or away from our observers of 10 cm/s, the Sun would chug along at a more substantial speed of 3,180 cm/s (or about 32 m/s). Given these numbers and the current limiting sensitivities of the instruments in astronomers' toolkits, while detecting an Earth-mass planet in an Earth-size orbit is still impossible with the radial velocity technique, astronomers can now easily detect Jupiter-like planets in Earth-size orbits (or in smaller or even somewhat bigger orbits).

Through the current time, the radial velocity technique has been the workhorse approach for discovering exoplanets, with more than 50 % of all the planets known at the end of 2013 having been discovered through the reflex motions of the parent stars. Two of the least massive exoplanets yet discovered, a 1.9 Earth-mass object orbiting the nearby star Gliese 581 (known as Gliese 581 e)[32] and a 1.1 Earth-mass object orbiting Alpha Centauri B (known as Alpha Centauri Bb)[33] were both discovered with this technique. At the other end of the mass spectrum, hundreds of super-Jupiters have been found with this technique. Radial velocity measurements have also revealed planets in orbits as quick as 46 h (Gliese 876 d has an orbit almost 20 times smaller than the orbit of Mercury)[34] and as slow as 38 years (47 Ursa Majoris d has an orbit larger than the orbit of Saturn).[35] In addition, many planetary systems have been found via this technique, including the 55 Cancri system, which has at least five planets ranging in size from half the mass of Saturn to four times the mass of Jupiter.[36]

THE TRANSIT TECHNIQUE

When the Moon passes in between the Earth and the Sun, it blocks our view of the Sun. Because the Moon, when seen from the Earth, happens to have nearly the same angular size as the Sun, the Moon sometimes blocks most of the Sun (a partial solar eclipse) or other times all of the Sun (a total solar eclipse). If we were measuring the amount of sunlight we received at each moment on the day of a total solar eclipse, we would find that the Sun was enormously bright from sunrise (let's call that 6 a.m.) until shortly before midday (say 11:00 a.m.). Then, as the solar eclipse begins, the total amount of light we received from the Sun would slowly start to decrease. Shortly after noon, for a period of only a few minutes, during the brief period of total eclipse, the amount of sunlight would drop to nearly zero. When the short period of total eclipse ended, the amount of sunlight would

[32] Mayor, M. et al. (2009). The HARPS search for southern extra-solar planets. XVIII. An Earth-mass planet in the GJ 581 planetary system. *Astronomy & Astrophysics, 507*, 487.

[33] Dumusque, X., et al. (2012). An Earth-mass planet orbiting α Centauri B. *Nature*, 491, 207.

[34] Rivera, E., et al. (2005). A ~7.5 $M_{\oplus}$ planet orbiting the nearby star, GJ 876. *Astrophysical Journal, 634*, 625.

[35] Gregory, P., & Fischer, D. (2010). A Bayesian periodogram finds evidence for three planets in 47 Ursae Majoris. *Monthly Notices of the Royal Astronomical Society, 403*, 731.

[36] *The exoplanets Encyclopedia*. Retrieved from http://exoplanet.eu/

gradually increase until shortly after 1 p.m., when the brightness of the Sun would return to normal levels. Astronomers would call a plot of the amount of sunlight received on Earth as a function of time a *lightcurve* for the Sun. This particular lightcurve would contain evidence for the existence of the Moon. The evidence would be the drop from and return to the normal brightness level for the Sun.

The planet Venus is much further from the Earth than is the Moon. For observers on the Earth Venus appears about 30 times smaller in angular diameter in the sky (and so covers an area of the sky about 900 times smaller) than does the Moon. As a result, Venus can never cause a total eclipse of the Sun; it can, however, pass in front of the Sun and block out a small amount of sunlight. When Venus passes directly in front of the Sun as seen from Earth, as it does twice approximately every 120 years (transits of Venus occurred in 1631 and again 8 years later in 1639, in 1761 and then in 1769, in 1874 and again in 1882, and in 2004 and again in 2012), Venus blocks about one-tenth of one percent (one thousandth) of the total amount of light the Earth normally would receive from the Sun; such transits last for about 6 h. If astronomers did not know that Venus existed, but if they had extremely accurate measurements of the brightness of the Sun for every hour of every day extending back to 1600, they would discover that eight times in 400 years the brightness of the Sun had dropped by 0.1 %, each time for about 6 h. With careful thought and analysis, they would be able to deduce from these rare and unusually patterned but nevertheless repeatable events the existence of a Venus-size planet (only 5 % smaller in diameter than the Earth) in a 225-day orbit around the Sun.

Astronomers now use this transit technique to search for exoplanets. By very accurately measuring the brightnesses of individual stars, minute after minute, hour after hour, night after night, year after year, many different teams of astronomers are looking for tiny changes in the individual brightnesses of millions of stars. If a star fades by one-half percent for only a few minutes every 18 h, then astronomers know that a planet orbits that star in 18 h. If a star fades by one-tenth of one percent every 300 days, they know a planet orbits that star every 300 days. In 1999, Greg Henry, of Tennessee State University, and his collaborators made the first successful identification of a planet using the transit technique when they confirmed the existence of a planet that had previously been discovered via the radial velocity technique.[37] That planet, HD 209458 b, has a mass of about 70 % the mass of Jupiter and orbits its parent star in only 3.5 days.

This technique of looking for planetary transits of their host stars is the focus of NASA's Kepler satellite mission, the European Space Agency's CoRoT (Convection, Rotation and Transits) satellite mission, and many ground-based projects, including SuperWASP (Wide-Angle Search for Planets; based in the Canary Islands and South Africa), HATNet (Hungarian Automated Telescope Network; based in Arizona and Hawaii), QES (the Qatar Exoplanet Survey, sited in New Mexico), KELT (the Kilodegree Extremely Little Telescope, with KELT-North sited in Arizona and KELT-South sited in South Africa), as well as many other research teams positioned all around the Earth.

In August 2009, a team led by Alain Leger of the Université Paris-Sud announced their discovery CoRoT-7b, a planet a bit bigger and a bit more massive than Earth that

---

[37] Henry, G. W., et al. (2000). A transiting '51 Peg-like' planet. *Astrophysical Journal, 529,* L41.

orbits a star in the constellation Monoceros that lies at a distance of about 150 parsecs from the Sun (a parsec is a distance equal to about 3.26 light years).[38] Leger's discovery of CoRoT-7b was an early example of the power of the transit technique for identifying planets comparable in size to the Earth rather than to Jupiter (Jupiter's radius is more than 11 times greater than the radius of the Earth). CoRoT-7b has a measured radius of 1.68 Earth radii (and a mass estimated to be comparable to or smaller than the mass of Neptune). It orbits in about 20.5 h at a distance of only 0.017 astronomical units (one astronomical unit is the distance from the Earth to the Sun, about 93 million miles or 150 million kilometers), which places it about 60 times closer to its host star than the Earth is to the Sun.

About 30 % of all planets discovered through the end of 2013 have been discovered or measured via transits (note that a few planets have been observed via both the radial velocity and transit techniques). Some transiting planets are tiny in mass while others are substantially more massive than Jupiter. One of the powerful benefits of transit measurements is that observers can measure the physical size of the planets from the light curve data, though that information does not directly yield the mass of the planet. If, however, astronomers can apply both the transit method and the radial velocity method to a single planetary system, the measurements yield not only the size and mass but the density of the planet. The greatest successes achieved via transit technique measurements thus far have been made by the Kepler satellite. A detailed description of most of the Kepler discoveries made through the end of 2013 is contained in the Appendix.

THE MICROLENSING TECHNIQUE

One of the many bizarre ideas that emerged from Albert Einstein's theory of relativity is the bending of starlight. These, at least, are the words we commonly use to describe what happens when light passes near a massive object like the Sun. Rather than follow what appears to observers to be a straight line as the light moves past that massive object, the light curves back towards the object as it passes by, tracing the shortest possible path allowed by the laws of physics through the curved space around the object.

Because light rays follow curved paths when they pass near massive objects, light rays that otherwise might have followed parallel paths past such an object are instead brought together, as if the massive object were a lens in a pair of eyeglasses or a magnifying glass. The light is focused such that a distant object appears much brighter that it would in the absence of the massive *gravitational lens*.

When a foreground star passes in front of a much more distant star, light from the distant star will be gravitationally lensed by the gravitational field of the foreground star (the foreground star is the gravitational lens). For the brief period of time required for the complete passage of a foreground star in front of a background star—typically about 50 days—astronomers would observe a brightening of the distant star. The brief period when the distant star appears brighter than its normal brightness is called a microlensing event.

Now let's add an additional complication: if the foreground star is orbited by a planet, the microlensing lightcurve will show an additional, much briefer (less than a day)

---

[38] Léger, A., et al. (2009). Transiting exoplanets from the CoRoT space mission. *Astronomy & Astrophysics, 506*, 287.

microlensing event caused by the planet when both the foreground star and planet simultaneously act as gravitational lenses.

The first planet detected via the microlensing technique is a planet more than twice as massive as Jupiter, detected in the microlensing event identified as OGLE 2003-BLG-235Lb. This planet was detected in 2003 by both the Optical Gravitational Lensing Experiment (OGLE) team and the Microlensing Observations in Astrophysics (MOA) team.[39] One year later, the OGLE team detected a planet of about four Jupiter masses in the microlensing event OGLE-2005-BLG-71Lb.[40] Microlensing is a successful technique for detecting planets—by the end of 2013, two dozen planets orbiting 22 different stars had been identified in this way. So far, the planetary masses of microlensed planets range from as small as about three Earth masses to as large as nine Jupiter masses (3,000 Earth masses).

Microlensing, however, suffers a major disadvantage in comparison to radial velocity or transit discoveries, because during a microlensing event we can identify the background (lensed) star, but we do not know the identity of the foreground (lensing) star around which the planet orbits. As a result, from microlensing measurements, we can learn about the statistical likelihood of planets, but astronomers will be unable to do any follow-up studies on the microlensed planets themselves. Microlensing surveys continue with observations being made with an upgraded MOA facility known as MOA-II, an upgraded OGLE telescope known as OGLE-IV, at the Wise Observatory in the Negev desert south of Tel Aviv, and with plans for a microlensing capability on the future NASA mission called WFIRST (the Wide-Field InfraRed Survey Telescope).

THE DIRECT IMAGING TECHNIQUE

See the planet. Take a picture of the planet. That's direct imaging.

Everybody wants pictures of the exoplanets as they are discovered, but very few of these exoplanets, so far, are susceptible to direct imaging. The principle problem for direct imaging is that the planets are faint objects that are very close to the extremely bright stars they orbit. If we were looking back at the Earth-Sun system from Alpha Centauri in the colors of light at which our eyes are most sensitive, the Sun would be almost ten billion times brighter than the Earth. The relative brightness of the Sun and faintness of the Earth would make the Earth virtually impossible to see unless we found a way to block out the direct light from the Sun without simultaneously obscuring the light from the Earth.

A group led by Gaël Chauvin, of the European Southern Observatory, used several novel ideas to overcome this factor-of-ten-billion problem that enabled them, in 2004, to successfully obtain the first image of an exoplanet.[41] First, they imaged the object 2M1207

---

[39] Bond, A., et al. (2004). OGLE 2003-BLG-235/MOA 2003-BLG-53: A planetary microlensing event. *Astrophysical Journal, 606,* L155.

[40] Udalski, A., et al. (2005). A Jovian-mass planet in microlensing event OGLE-2005-BLG-071. *Astrophysical Journal, 628,* L109.

[41] Chauvin, G., et al. (2004). A giant planet candidate near a young brown dwarf. *Astronomy & Astrophysics, 425,* L29.

at infrared wavelengths rather than in visible light. In the infrared, the planet *emits* far more light of its own than the amount of starlight it reflects, whereas in visible light the planet merely acts as a dirty mirror and *reflects* a small amount of starlight while emitting almost no light at all. In addition, the star is fainter in the infrared than at visible wavelengths. This combination—the planet is brighter and the star is fainter in infrared light, as compared to visible light—yields a good strategy for direct imaging. Finally, rather than looking for a planet in orbit around a star, they looked for a planet in orbit around a brown dwarf. Because brown dwarfs are intermediate in mass between stars and planets, when they are young they are much brighter than planets but much fainter than stars. This combination creates a better contrast ratio between the star and the planet. The brown dwarf 2M1207 is about 25 times more massive than Jupiter and is about 500 times fainter than the Sun. The planet found by Chauvin and his team, 2M1207 b, has a mass of about four Jupiters.

The next big success in imaging exoplanets was achieved nearly simultaneously in 2008 by two groups, one led by Paul Kalas, of the University of California at Berkeley, and the other led by Christian Marois, of the NRC Herzberg Institute of Astrophysics in Canada. Kalas's group imaged a less-than-three Jupiter mass planet in orbit around the nearby star Fomalhaut[42] while Marois's team imaged three slightly more massive planets in orbit around the more distant star HR 8799.[43] Two years later, another team led by Marois imaged a fourth planet around HR 8799.[44]

Another super-Jupiter, or perhaps a small brown dwarf, was imaged by a team led by Phillipe Delorme of Joseph Fourier University in Grenoble. Using data obtained at the Very Large Telescope in Chile in 2002 and 2012, they have both imaged and measured the orbital motion of the planet identified as 2MASS0103(AB)-b. This object, whose mass is probably at the very upper limit (or above) for planets, in the 12–14 Jupiter-mass range, orbits at a distance of 84 astronomical units from a pair of young (30 million years old), low mass stars. This star-planet system is relatively nearby, only 47 parsecs from the Sun.[45]

In late 2013, a team lead by Michael Liu of the Institute for Astronomy at the University of Hawaii announced they had discovered, through the technique of direct imaging, a super-planet, one 6 times more massive than Jupiter. The planet known as PSO J318.5–22 was found drifting through space on its own, that is, not in orbit around a parent star.[46] Such free-floating, or rogue, planets might be rare. And they may be irrelevant when we think about locations where life could exist, since without heat and light from a nearby star, this planet is likely far too cold to support life. But the discovery does demonstrate

[42] Kalas, P., et al. (2008). Optical images of an exosolar planet 25 light-years from earth. *Science, 322*, 1345.

[43] Marois, C., et al. (2008). Direct imaging of multiple planets orbiting the star HR 8799. *Science, 322*, 1348.

[44] Marois, C., et al. (2010). Images of a fourth planet orbiting HR 8799. *Nature, 468*, 1080.

[45] Delorme, P., et al. (2013). Direct Imaging discovery of 12-14 Jupiter mass object orbiting a young binary system of very low mass stars. *Astronomy & Astrophysics, 553*, L5.

[46] Liu, M. C., et al. (2013). The extremely red, young L dwarf PSO J318.5338-22.8603: A free-floating planetary mass analog to directly imaged young gas-giant planets. *Astrophysical Journal, 777*, L120.

that astronomers are rapidly developing the capabilities to direct image planets. By the end of 2013, claims had been made—some of them disputed—for the detection through direct imaging of about one dozen planets and about two dozen brown dwarfs orbiting a total of 38 stars or brown dwarfs.

TIMING TECHNIQUES

Several different techniques, all of which are known as timing techniques, have been used to identify several dozen planets (see Appendix for the Kepler-related discoveries), as of the end of 2013. These timing techniques measure tiny temporal deviations from the regularity of other periodic events.

Pulsar Timing Variations. Pulsars (neutron stars) are the remnants of explosions that destroy massive stars when they run out of fuel and die. A typical pulsar has two to three times the mass of the Sun, is only a few kilometers in diameter and spins hundreds of times or even as fast as a 1,000 times per second. Pulsars have very strong magnetic fields at their surfaces. The intense magnetic field of a pulsar causes most of the light from such an objects to be emitted outwards along the direction of the magnetic axis, and from both the north and south magnetic poles. Like the Earth, the magnetic axis typically is not perfectly aligned with the rotation axis of a pulsar. As a result, as the pulsar spins, the beams of light emerging from the magnetic poles sweep around in circles, like beams of light from a lighthouse. If the Earth happens to lie in a direction that is swept by the light beam of the pulsar, astronomers see a source of light that appears to pulse on and off, although all that is actually happening is that the beam of light from the pulsar is sweeping into and out of our line of sight; the appearance of the beam of light turning on and off is the reason that neutron stars are also called pulsars. Amazingly, the rapid spins of pulsars make them extremely precise astrophysical clocks, with the regularity of the flashes of light from pulsars comparable to the dependability of the ticking of terrestrial atomic clocks.

Now, imagine a pulsar that spins nearly 1,000 times per second and that has company. This pulsar is in a system in which it orbits another object. Consequently, the beam of light from this pulsar points toward and then away from the Earth 1,000 times per second, and astronomers would measure 1,000 pulsar pulses per second from this pulsar. Such an object is known as a millisecond pulsar. As this millisecond pulsar orbits its companion, it alternately moves toward and then away from the Earth, generating a Doppler shift in the light emitted toward the Earth. That toward and away-from movement of the pulsar generates extremely tiny variations in the millisecond ticks of the pulsar. The sizes of those pulsar-timing variations (seconds, tenths of seconds, or hundredths of seconds) contain information about the mass of the object around which the pulsar orbits. In a few cases, astronomers have determined that these timing variations are caused by planets rather than by other stars.

The planets discovered via this technique by Wolszczan and Frail around PSR 1257 + 12 in 1991 were the first three of five pulsar planets now known; the initial discoveries, however, have not led to an avalanche of discoveries of pulsar planets. Pulsar PSR

1719–14 has a Jupiter-sized planet and PSR B1620–26 has a planet two and one-half times larger than Jupiter,[47,48] while PSR 1257 + 12 is known to have three planets.

Eclipse Timing Variations. A handful of planets have been discovered through the timing of eclipses. In a binary star system in which the two stars periodically pass in front of and behind each other, the eclipses occur with very precise regularity. If, however, a third object, a planet, exists in such a system that planet will cause regular changes in the timing of the eclipses. Eclipse timing events have been used to discover planets in a small handful of systems, including a super-Jupiter around UZ For,[49] a super-Jupiter around HU Aqr,[50] two super-Jupiters around NN Ser,[51] one super-Jupiter around NY Vir,[52] one likely brown dwarf around HW Vir,[53] and one super-Jupiter around DP Leo.[54]

Transit Timing Variations. The presence of multiple planets orbiting a single star offers an additional physical effect astronomers have exploited to find planets. If the planets are massive enough or have orbits that are similar enough, their gravitational pulls *on each other* will affect their respective orbits. Earth, for example, tugs on Venus and causes the orbital period of Venus (224.7 days) to vary by about 10 min.[55] This method has become a workhorse tool for retrieving planets from the Kepler database (see Appendix for details of these discoveries).

Pulsating Star Timing Variations. When stars that are much more massive than the Sun reach old age, they puff their outer layers off into space. In rare circumstances, the core of the star that is left behind becomes unstable and begins to pulsate, alternately expanding and contracting (these stars are pulsating stars, but are not pulsars). The change in size of the core forces the entire star to change in size. As the star becomes bigger it becomes brighter; then, as it decreases in size it becomes fainter. Astronomers are not able to directly measure the sizes of these stars, but they have no problem measuring the regular

---

[47] Bailes, M., et al. (2011). Transformation of a star into a planet in a millisecond pulsar binary. *Science, 333*, 1717.

[48] Rasio, F. A. (1994). Is there a planet in the PSR 1620-26 triple system? *Astrophysical Journal, 427*, L107.

[49] Dai, Z., et al. (2010). Orbital period analyses for two cataclysmic variables: UZ Fornacis and V348 Puppis inside the period gap. *Monthly Notices of the Royal Astronomical Society, 409*, 1195.

[50] Qian, S.-B., et al. (2011). Detection of a planetary system orbiting the eclipsing polar HU Aqr. *Monthly Notices of the Royal Astronomical Society, 414*, L16; Goździewski, K., et al. (2012). On the HU Aquarii planetary system hypothesis. *Monthly Notices of the Royal Astronomical Society, 425*, 930.

[51] Beuermann, K., et al. (2010). Two planets orbiting the recently formed post-common envelope binary NN Serpentis. *Astronomy & Astrophysics, 521*, L60.

[52] Qian, S.-B., et al. (2012). Circumbinary planets orbiting the rapidly pulsating subdwarf B-type binary NY Vir. *Astrophysical Journal, 745*, L23.

[53] Lee, J. W., et al. (2009). The sdB + M eclipsing system HW virginis and its circumbinary planets. *Astronomical Journal, 137*, 3181.

[54] Qian, S.-B., et al. (2010). Detection of a giant extrasolar planet orbiting the eclipsing polar DP Leo. *Astrophysical Journal, 708*, L66.

[55] Murray, N. W. (2012). Evidence of things not seen. *Science, 336*, 1121.

changes in brightness of pulsating stars. These pulsation periods can be timed extremely accurately. As with pulsars and eclipsing binaries, a planet orbiting a pulsating star will subtly and regularly modulate the pulsation period. The mass of the planet can be calculated from the effect it has on the pulsation period.

The star V391 Pegasi is one such star. After shedding 50 % of its mass, the hot stellar core began to pulsate, with a pulsation period of about 350 s. Such stars, known as subdwarf B pulsators ('subdwarf' stars have physical sizes smaller than those of regular stars; 'B' indicates that the surface temperature of this star is 2 to 6 times hotter than the Sun), have extremely stable pulsation periods. In the case of V391 Pegasi, the research team led by R. Silvotti, of the Osservatorio Astronomico di Capodimonte, in Italy, discovered a very regular 3.2-year variation in the 350-s pulsation period.[56] They concluded that the period of variability of 3.2 years is caused by a planet about 3 times more massive than Jupiter orbiting the star at a distance comparable to the distance of Mars from the Sun. Using the same technique, S.-B. Qian, of the National Astronomical Observatories in China, and collaborators report the presence of a Jupiter-size planet in an 8-year orbit around the subdwarf B pulsator star NY Vir.[57]

ASTROMETRY TECHNIQUE

Imagine you run a security company that takes satellite reconnaissance photographs every minute, all night long, night after night, of a parking lot full of cars parked by drivers who are now out of town on a month-long cruise. Since the cars are parked and without drivers, minute after minute, night after night the cars don't move. All the cars should maintain their positions relative to all the other cars. One night, before any of the drivers have returned from their vacation, you notice one car start to change positions relative to all of the other cars. You report a probable stolen vehicle attempt in progress.

Now imagine that night after night you photograph the sky in the vicinity of your favorite star. Your images capture your target star in addition to thousands of other stars. From your photographs, you measure the relative positions of all the stars. You find that all but two of the other stars never move relative to each other; all but these two stars are fixed in their positions. These two stars, however, continually change their positions relative to the positions of the other, fixed stars. Slowly but surely, these two stars trace out elliptical paths around their common center of mass. Through your measurements of the changing positions of these two stars, i.e., using the technique called astrometry, you have discovered a binary star system. Beginning with observations made more than 200 years ago, astronomers have discovered many binary star systems in this way.

Now imagine that one of the two stars in the binary system is very faint in comparison to its partner. Maybe it's a white dwarf or a neutron star or even a black hole. It might even

---

[56] Silvotti, R., et al. (2007). A giant planet orbiting the 'extreme horizontal branch' star 391 Pegasi. *Nature, 449*, 189.

[57] Qian, S.-B., et al. (2012). Circumbinary planets orbiting the rapidly pulsating subdwarf B-type binary NY Vir. *Astrophysical Journal, 745*, L23.

be too faint for astronomers on Earth to see it. In your observations, you see one star move round and round while all the other stars remain at their fixed positions. You deduce that the one star you can see is orbiting an unseen companion. From the size of the orbit, the orbital period, the mass of the star you can see and the law of gravity, you can calculate the mass of the unseen binary companion star. This technique, in fact, was used more than a century ago to discover the first white dwarf star, Sirius B, in orbit around Sirius (Sirius B has the mass of the Sun and the diameter of the Earth; because of its small size, it is exceedingly faint in comparison to Sirius); this astrometry technique was also used in the late twentieth century to discover small black holes, such as Cygnus X-1 (with a mass of about 11 Suns), and even the supermassive black hole at the center of the Milky Way.

Peter van de Kamp used the astrometry method in his flawed studies of Barnard's Star. The actual motions of a star due to its orbital motion around a planet would be thousands, even hundreds of thousands, of times smaller than the astrometric motion caused by a star orbiting another star (or orbiting a white dwarf or black hole). Such measurements are exceedingly difficult to make.

For a star orbited by a planet, the observable astrometric signature for the motion of the star would be a periodic change in position of about one thousandth of one second of arc (1 milli-arcsecond) or smaller. For comparison, the full moon spans an angular size of one half of one degree. One half degree is equal to 30 min of arc, or 1,800 seconds of arc, or 1,800,000 milli-arcseconds. Imagine, then, slicing the full moon into 1,800,000 slices of equal width. Your goal, as an astrometrist, is to measure positional changes of stars equal to a fraction of the angular width of one of these slices.

In 2009, Steven Pravdo and Stuart Shaklan of the Jet Propulsion Laboratory published a paper in which they claimed to have succeeded where van de Kamp had failed: they claimed to have made the first planet discovery using the technique of astrometry.[58] According to their measurements, they found a planet 6 times more massive than Jupiter in a 0.744 year orbit around the star VB 10. This planet, however, like van de Kamp's planets around Barnard's Star, has not withstood the test of time or the challenge of scientific reproducibility. Only 9 months after the VB 10 planet was 'discovered,' a team led by Jacob L. Bean, of the Institute für Astrophysik in Germany, used the radial velocity technique to try to confirm the reality of VB 10b. The title of their paper is very direct: "The Proposed Giant Planet Orbiting VB 10 Does Not Exist." [59] Other research groups independently refuted the original claim, using both astrometry[60] and radial velocity[61] measurements. In 2010, the Palomar High-precision Astrometric Search for Exoplanet Systems (PHASES) team, led by Matthew Muterspaugh of Tennessee State University, announced "with some trepidation" astrometric evidence for a 1.5 Jupiter-mass superJupiter in a

[58] Pravdo, S., & Shaklan, S. (2009). An ultracool star's candidate planet. *Astrophysical Journal, 700*, 623.

[59] Bean, J. L. (2010). The proposed giant planet orbiting VB 10 does not exist. *Astrophysical Journal Letters, 711*, L19.

[60] Lazorenko, P. F., et al. (2011). Astrometric search for a planet around VB 10. *Astronomy & Astrophysics, 527*, A25.

[61] Anglada-Escudé, G., et al. (2010). Strong constraints to the putative planet candidate around VB 10 using Doppler spectroscopy. *Astrophysical Journal Letters, 711*, L24.

1,016-day orbit around HD 176051. The PHASES team writes that they have 'high confidence' that their measurements are real changes in the astrometric position of HD 176051 rather than any kind of error in their work.[62] So far, their work has been neither confirmed nor refuted.

We are still waiting for the first definitive and confirmed discovery of a planet via the astrometric technique. To this end, several research teams are now actively searching for planets with instruments that in principle can achieve sub-milli-arcsecond accuracies. These include ASPENS (Astrometric Search for Planets Encircling Nearby Stars), RIPL (the Radio Interferometric Planet search) and CAPS (the Carnegie Astrometric Planet Search). In addition, the European Space Agency launched the Gaia mission on 19 December 2013. The Gaia satellite, which will orbit the Sun at a position 1.5 million kilometers from the Earth in the opposite direction from the Sun, a location known as the second Lagrange point (L2), will house a telescope that has been designed to measure the positions of one billion stars at an accuracy of 24 microarcseconds (millionths of a second of arc) for five and one-half years. Gaia is expected to detect tens of thousands of planets, some through transits, others through radial velocity measurements, and some via astrometry. Perhaps one of these project teams will succeed in using astrometry to detect an exoplanet.

HABITABLE ZONE PLANETS

Life might be able to take root in environments unimaginable to human scientists. Currently, however, serious conversations about extraterrestrial life focus on planets and moons in regions around stars known as habitable zones. These planets are the holy grail of exoplanet studies.

The habitable zone around any single star is loosely defined as the region around a star in which orbiting rocky planets (or large moons with solid surfaces) with atmospheres dominantly composed of carbon dioxide, water and molecular nitrogen would receive enough light from their central stars to be able to have liquid water on their solid surfaces at some time during the lifetime of a star. Because stars become a few percent brighter as they age, the habitable zone moves outward from the star a little bit as stars grow older.

Astronomers have at last broken the barrier in finding planets in or near habitable zones. A few of these planets share some properties (size, mass, temperature or density) with the Earth. Over the coming decades, an enormous number of habitable zone planets likely will be discovered. Those habitable-zone planets already discovered include the ones known as Kepler-22b, Kepler-62e, Kepler-62f, Kepler-69c, Kepler-61b, and Kepler-47c. In each case, the star designated by the name 'Kepler' combined with a number is part of the group of stars surveyed by NASA's orbiting Kepler telescope. The letters b, c, e and f indicate that the planets discovered were the second (b), third (c), fifth (e) and sixth (f)

[62] Muterspaugh, M. W., et al. (2010). The PHASES differential photometry data archive. V. Candidate substellar companions to binary systems. *Astronomical Journal, 140*, 1657.

objects known in each stellar system. The star itself, usually identified without using the letter a, is the first known object in each system.

The remainder of this section details chronologically the discoveries of these five planets that, by virtue of being habitable-zone planets, are especially relevant to questions about extraterrestrial life that might be similar in at least some respects to life on Earth.

In 2012, William Borucki and his team of Kepler scientists identified the first-known nearly Earth-like planet, Kepler-22b, in the habitable zone of a star other than the Sun.[63] The star Kepler-22 is extremely Sun-like in size (98 % the radius of the Sun), mass (97 % the mass of the Sun) and brightness (79 % the luminosity of the Sun). The size of the orbit of Kepler-22b (0.85 astronomical units) would place it halfway in between Venus and Earth, were it in orbit around the Sun. As a result of the size of the orbit and the temperature and brightness of the star, Kepler-22b has an "equilibrium temperature" of 262 K, which is just below the freezing point (273 K) and well below the boiling point (373 K) for water. The calculated equilibrium temperature assumes the planet has no atmosphere (the atmosphereless Moon is a good example of an object whose surface temperature is the same as its equilibrium temperature). A planet with an atmosphere will be somewhat warmer than the equilibrium temperature, because the atmosphere acts like a greenhouse. The Earth's atmosphere, for example, raises the average temperature of the Earth by 33 K. Most likely, the temperature of Kepler 22b is above freezing, though future observations will be required to test that hypothesis. Kepler-22b is a bit larger than the Earth (2.4 Earth radii) and is likely more like Uranus and Neptune than like Earth in composition and structure. Its mass is not well known; we only know that its mass is very likely less than the mass of Saturn. Even if Kepler-22b itself is uninhabitable, it could have Earth-size, Earth-like moons which would be in the habitable zone of Kepler-22.

In May 2013, a Borucki-led team found five planets orbiting the star Kepler-62.[64] The five-planet system Kepler-62 includes two super-Earths in the habitable zone of Kepler-62 and three other small planets too close to the star to be in the habitable zone. Kepler-62b, Kepler-62c and Kepler-62d orbit in 5.71 days, 12.44 days and 18.16 days, have radii of 1.31, 0.54 (Mars-sized) and 1.95 Earth radii, and have unknown masses that are estimated to be no greater than about 9, 4 and 14 Earths, respectively. The most intriguing planets found by Borucki orbiting Kepler-62 are Kepler-62e and Kepler-62f, which orbit the central star in 122.4 days and 267.3 days, respectively. Both planets are super-Earths, with radii of 1.61 and 1.41 Earth radii and with estimated masses of up to but no greater than 36 and 35 Earth masses. If they are at the top range of these mass limits, they are super-Neptunes, but super-Neptunes could have Earth-size moons, and these planets or those possible moons would be at distances from Kepler-62 such that their equilibrium temperatures would be 270 and 208 K, which (though below the freezing point of water today, assuming no atmospheric warming) would place them well within the habitable zone around the central star, at some time during the lifetime of the star.

---

[63] Borucki, W. J., et al. (2012). Kepler-22b: A 2.4 Earth-radius planet in the habitable zone of a Sun-like star. *Astrophysical Journal, 745*, 120.

[64] Borucki, W. J., et al. (2013). Kepler-62: A five-planet system with planets of 1.4 and 1.6 earth radii in the habitable zone. *Science, 340*, 578.

Also in May 2013, Thomas Barclay, of NASA's Ames Research Center, and his team found a super-Earth-size planet in the habitable zone of Kepler-69.[65] They also found a second planet close to the star. As Kepler-69 is very Sun-like (81 % of the mass of the Sun, 80 % the luminosity of the Sun, almost the same surface temperature as the Sun), this system has one of the closest analogues to the Earth yet found. Kepler-69b is a Uranus-like planet (2.24 Earth radii) in a 13.7 day orbit while Kepler-69c is a super-Earth (1.7 Earth radii) in a 242.5 day orbit. Barclay calculates that if the amount of starlight reflected by Kepler-69c is similar to the amount of sunlight reflected by the Earth, Kepler-69c would have a temperature of 300 K, which is hot by terrestrial standards but well above the freezing point and well below the boiling point of water; however, Stephen Kane, of the Exoplanet Science Institute at Cal Tech, argues that Kepler-69c is much more Venus-like than Earth-like.[66] Venus orbits the Sun in 224.7 days and receives 1.91 times more sunlight per square meter of surface than does the Earth; Kepler-69c, in a very similar orbit to Venus and in an orbit around a very similar star to the Sun, also receives about 1.9 times more light from its star as the Earth receives from the Sun. Venus is in the habitable zone of the Sun but is no longer habitable. Venus almost certainly once had liquid water and probably large oceans on its surface; however, a billion years ago or thereabouts, the greenhouse effect on Venus grew more and more intense. Slowly, a runaway greenhouse heated the atmosphere of Venus by 500 degrees. Over time, the atmosphere became so hot that the oceans boiled off the surface of Venus. Ultimately, most of the hydrogen atoms that were bonded to oxygen atoms in the water molecules were lost to space. The surface of Venus is now 737 K, about 360 K above the boiling point for water. If Kepler-69c is like Venus, it may no longer be habitable, but it might have been at one time in its history.

Kepler-61b is also a habitable-zone planet. Sarah Ballard, of the University of Washington, and her team found Kepler-61b in a 59.88-day orbit around a star that is less massive, less luminous and cooler than the Sun.[67] With a radius of 2.15 Earth radii, Kepler-61b is either a super-Earth or a mini-Neptune. Ballard estimates the mass to be about 3.2 Earth masses, though that number, for now, is merely an educated guess. The actual mass could easily be several times greater. While the temperature of the central star in this planetary system is barely two-thirds that of the Sun, 4,000 K versus 5,780 K, Kepler-61b is four times closer to Kepler-61 (0.26 astronomical units) than the Earth is to the Sun. As a result, the temperature at the top of the atmosphere of Kepler-61b should be 273 K, which is close to the current, globally-averaged surface temperature (including the current level of greenhouse-gas heating) of the Earth, 288 K. This planet will be of great interest for future observations, because this system is more than 1 billion years old. When the Earth was only 1 billion years old (3.5 billion years ago), life had already taken hold in the form

[65] Barclay, T., et al. (2013). A super-Earth-size planet orbiting in or near the habitable zone around a Sun-like star. *Astrophysical Journal, 768*, 101.

[66] Kane, S. R., et al. (2013). A potential super-Venus in the Kepler-69 system. *Astrophysical Journal Letters, 770*, L20.

[67] Ballard, S., et al. (2013). Exoplanet characterization by proxy: A transiting 2.15 $R_\oplus$ planet near the habitable zone of the late K Dwarf Kepler-61. *Astrophysical Journal, 773*, 98.

of complex, mat-forming microbial communities known as stromatolites,[68] and even if Kepler-61b is a mini-Neptune, it could have an Earth-size moon that could have an Earth-like composition and an atmosphere like that of the young Earth.

Kepler-47, nicknamed the Tatooine system after the twin stars in the sky above Luke Skywalker's home planet in the Star Wars saga, is a binary star system with two Neptune-size planets orbiting the central binary. The central binary is unusual because it has two mismatched stars—Kepler-47A is like the Sun while Kepler-47B has a mass of about one-third that of the Sun. The stars orbit each other in 7.45 days. Jerome Orosz, of San Diego State University, and his team found Kepler-47b in a 49.5-day orbit and Kepler-47c in a 303-day orbit.[69] Kepler-47b is about three times bigger in radius than the Earth; its mass is unknown but it must be no more massive than two Jupiters, otherwise its mass would affect the orbits of the stars in ways that would have been detected in the Kepler data. The orbit of Kepler-47c places it at 0.99 astronomical units from the center of this complicated system, that is, almost exactly the same distance as the Earth is from the Sun. Thus, Kepler-47c is squarely in the habitable zone. Though Kepler-47c is not Earth-like—it is slightly larger than Uranus, at 4.6 Earth radii, and has a mass that is probably in the range of 16–23 Earth masses, comparable to that of Neptune—it could have Earth-size and thus Earth-like moons which would be in the habitable zone, which makes it of great interest for further study.

STATISTICS: HOW MANY HABITABLE ZONE PLANETS MAY EXIST IN THE MILKY WAY

In October 2013, Erik A. Petigura, then a graduate student at the University of California at Berkeley, led a team of astronomers that asked the question "How many Sun-like stars in the Kepler study sample show evidence of having Earth-size planets in Earth-like orbits?" They found that about 42,000 of the 150,000 stars studied by Kepler were Sun-like and that 603 of the 42,000 showed evidence in the Kepler data for planets.[70] Of the 603 planet candidates, 10 "are Earth size and orbit in the habitable zone, where conditions permit surface liquid water."

Ten sounds like a small number of Earth-size planets discovered in the habitable zone out of the 42,000 stars in the sample. We need to remember, however, that Kepler can only detect a planet if the orbit of the planet brings it directly in front of the host star. If the orbital plane of the planet is tilted just a little bit, the planet would pass just above or just below our direct line-of-sight to the star, but the planet would never transit and dim the light of the host star. For these star-planet systems, Kepler would be unable to detect the presence of the planet. The questions that Petigura had to wrestle with are "What

---

[68] Noffke, N., et al. (2013). Microbially induced sedimentary structures recording an ancient ecosystem in the ca. 3.48 billion-year-old dresser formation, Pilbara, Western Australia. *Astrobiology, 13,* 1.

[69] Orosz, J. A., et al. (2012). Kepler-47: A transiting circumbinary multiplanet system. *Science, 337,* 1511.

[70] Petigura, E. A., Howard, A. W., & Marcy, G. W. (2013). Prevalence of Earth-size planets orbiting Sun-like stars. *Proceedings of the National Academy of Sciences United States of America, 110,* 19273.

percentage of planets could have been detected?" and "What fraction of the stars studied by Kepler could have planets that could not be detected?"

Big planets in tiny orbits are the easiest ones to detect. Because they are big, they block more starlight. Because they are close to their stars they are more likely to pass directly in front of the stars and not slip into just-above or just-below the star orbits. Statistically, about 8 % of Jupiter-like planets in quick (5 days or less) orbits should be detectable in transiting orbits. The other 92 % of the so-called hot Jupiters should escape detection.

What about planets with longer orbits? As the size of the orbit (and thus the orbital 'year' for the planet) increases, the probability that our viewing position is properly aligned with the orbit of a planet such that we would be likely to observe a transit event decreases very rapidly. For planets whose orbital periods are about 365 days, which would place these planets at about the same distances from their stars as the Earth is from the Sun, the transit probability for a Jupiter-like planet would be only about one-half of one percent. These probabilities mean that for every one such planet that we detect another 199 planets exist but are not detectable. In other words, we should detect only one-half of one percent of the Jupiters in Earth-like orbits.

What about smaller planets? The detection probability for a planet ten times smaller than Jupiter (i.e., an Earth-size planet) would be ten times smaller than the detection probability for a Jupiter-size planet. These numbers mean that for an Earth-size planet in an Earth-size orbit, the chances that Kepler would detect the planet are one-twentieth of one percent rather than one-half of one percent. Thus, for every one Earth-size planet in a 365-day orbit that Kepler should be able to discover, about 2,000 such planets exist that we cannot detect. Another way of thinking about these numbers is to say that every Earth-size planet in a 365-day orbit detected by Kepler in this 42,000 star sample represents 2,000 total Earth-size planets in 365-day orbits among those same 42,000 stars.

In addition, the quality of the brightness measurements made by Kepler is not uniformly excellent for all the stars; all other things being equal, the transit signal is weaker if the star is fainter.

Finally, the signature of the transit in the brightness profile of the star is harder to pick out if the star flickers. In this context, flickering refers to variations in the brightness of the star that occur naturally due to dynamic processes on the surface of the star and that could be at or even below the level of one-hundredth of one-percent of the star's luminosity. That level of flickering could be comparable to the brightness changes produced by a planetary transit. As a result, for a great many of the stars observed by Kepler, the intrinsic variations in the brightness of the star likely would obscure any possible dimming of the light of the star that results from the transit of the planet.

Based on considerations like these, Petigura calculates that the 10 Earth-size planets in habitable zones detected by Kepler allow us to conclude that "22 % of Sun-like stars harbor Earth-size planets orbiting in their habitable zones" and that "small planets far outnumber large ones." Given that the Milky Way includes about four hundred billion stars, these results suggest that at least several tens of billions of planets exist in the Milky Way and that a substantial fraction of these are Earth-size planets in Earth-like orbits. A fair number of these Earth-size planets in habitable zones should be in our neighborhood, orbiting stars within 50 light years of the Sun. Some of them might turn out to be not just Earth-size but Earth-like.

While this statistical extrapolation from 10 Earth-size planets in habitable zones out of 42,000 stars observed to several billion Earth-size planets in habitable zones out of the few hundred billion stars that exist in the Milky Way might appear rather bold, the important conclusion to draw from this work is exactly the opposite: the Kepler data have made clear what most astronomers have suspected as true for centuries—Earth-size planets are common in the Milky Way. Almost certainly, these results suggest that Earth-like planets are likely common as well.

THE NEXT GENERATION OF EXOPLANET TELESCOPE PROJECTS

The JWST (James Webb Space Telescope), the successor telescope to the Hubble Space Telescope (HST), is currently scheduled for launch in 2018. With its 6.5-m diameter mirror and the sensitive instruments being built specifically for this telescope, the JWST is designed to be able to obtain infrared images of giant planets and to measure their spectra. Already, astronomers are using the HST to test and demonstrate the kinds of science that will be doable with the JWST. Using the detector system known as the Wide Field Camera 3 on the HST, Drake Deming, Avi Mandell, H. R. Wakeford and their teams of collaborators observed six planets. All six of these planets had been discovered previously as they transited in front of their host stars. In the spectra of the six hot-Jupiters known as HD 290458b, XO-1b, WASP-12b, WASP-17b, WASP-19b and HAT-P-1b, which they obtained during transits of the planets in front of their respective host stars, they detected atmospheric water.[71] Future observations with JWST and other telescopes undoubtedly will do even better, detecting many other molecules, including potential biomarkers, in the atmospheres of exoplanets.

NASA selected the TESS (Transiting Exoplanet Survey Satellite) mission in April 2013 for development and a planned launch as a space-platform telescope in 2017. Rather than point in only one direction in the sky, as did Kepler, TESS will conduct a survey of the entire sky over a time period of 2 years. TESS will look for evidence for transits of Earth-size planets in habitable zones around 500,000 of the nearest and brightest stars in the sky. TESS team scientists predict that TESS will identify at least 2,000 planets.

The CHEOPS (Characterizing Exoplanet Satellite) mission was selected by the European Space Agency (ESA) in October 2012 for further study. CHEOPS will study transits of bright stars already known to host exoplanets, in order to measure the radii of the known exoplanets with an accuracy of 10 %. In combination with known masses of

---

[71] Deming, D., et al. (2013). Infrared transmission spectroscopy of the exoplanets HS 209458b and XO-1b using the wide field camera-3 on the Hubble space telescope. *The Astrophysical Journal, 774*, 95; Mandell, A. M., et al. (2013). Exoplanet transit spectroscopy using WFC3: WASP-12b, WASP-17b, and WASP-19b. *The Astrophysical Journal, 779*, 128; Wakeford, H. R., et al. (2013). HST hot Jupiter transmission spectral survey: Detection of water in HAT-P-1b from WFC3 near-IR spatial scan observations. *Monthly Notices of the Royal Astronomical Society, 435*, 3481.

these exoplanets, the data obtained with CHEOPS will allow astronomers to more accurately determine the densities, and thus the compositions, of these exoplanets.

The WFIRST (Wide-Field Infrared Survey Telescope) mission remains under study by NASA. WFIRST would utilize a 2.4-m telescope mirror donated to NASA by the National Reconnaissance Office in 2012. If developed and launched, plans for WFIRST include imaging of ice-giant and gas-giant exoplanets and microlensing searches to discover new planets.

The PLATO 2.0 (Planetary Transits and Oscillations of Stars) mission is under study by ESA. PLATO 2.0 is being designed to detect planets, including Earth-size planets in habitable zones; to measure transits in order to accurately measure the radii of exoplanets; to make radial velocity measurements in order to accurately measure masses of known Earth-size exoplanets; to provide accurate ages for the host stars and thereby for the planetary systems; and to identify target exoplanets for future spectroscopic studies of exoplanet atmospheres.

On the ground, telescopes keep getting bigger. Astronomers and engineers are currently designing both a 30-m telescope (the TMT) and a 39-m telescope (European Extremely Large Telescope), both of which could begin operation within two decades. These super-large telescopes will be able to measure spectra of the atmospheres of terrestrial-size exoplanets and detect more planets through multiple observing techniques.

PREDICTING THE FUTURE

Astronomers had discovered about 50 exoplanets by 2000. That number tripled to about 150 known exoplanets in 2005. It nearly tripled again, to about 400 in 2010. As of September 2011, astronomers had discovered 685 exoplanets around more than 500 other stars. By November 2012, the number of known exoplanets had climbed to 846, including 665 unique planetary systems and 126 different multiple-planet systems, some with as many as six planets. By December 2013, the numbers had reached 1,056 exoplanets in 802 planetary systems. [72] During the time period from 2010 to 2015, we can reasonably expect the number of known planets to once again have tripled, from about 400 to more than 1,200.

While the rate of increase of known exoplanets—tripling every 5 years—may not prove to be as dependable as Moore's Law, the 1965 assertion by the co-founder of Intel, Gordon Moore, that the number of transistors that engineers could place on a single integrated circuit would double every 2 years, we have no reason to think otherwise. By now, finding exoplanets has become one of the biggest projects worldwide for astronomers. The Extrasolar Planets Encyclopedia website, one of the most widely used global resources for information about exoplanets, identifies more than 100 ground-based and several dozen space-based exoplanet research teams and projects active or in the planning stages right now.

Given the amount of human, financial and telescopic resources astronomers are investing in working on this problem, the idea that the discovery rate of exoplanets is likely to continue, at minimum, to triple every 5 years for the foreseeable future is reasonable. At

---

[72] *The exoplanets Encyclopedia.* Retrieved from http://exoplanet.eu/

this pace of discovery, we might expect to know of more than 3,500 exoplanets by 2020, well over 10,000 exoplanets by 2025, approaching 30,000 exoplanets by 2030, an astounding one hundred thousand exoplanets by 2035, one-third of a million exoplanets planets by 2045, and a cool million exoplanets by 2045. By the end of the twenty-first century, catalogs of exoplanets are likely to include millions, if not tens of millions, of objects. We can quite reasonably expect that the number of known exoplanets will soon become, like the stars themselves, almost uncountable. In only a generation or two, we will reach the point of knowing, not merely surmising, that virtually every star has one or more planets.

We can therefore expect that within a single human lifetime, we will have populated the known sky with millions of planets. For those of us alive today, we can reasonably expect that within the lifetimes of our grandchildren, that number will explode to tens of millions of planets. Careful study of these many planets will open our eyes and minds to an aspect of the universe—life—about which we heretofore could only speculate.

If even only one of these planets shows evidence for biological activity, we will know that the Earth is not the only place in the universe where life exists. Perhaps all we will find is a planet with moss. But if moss can take root, advanced life forms could exist too. Just how many exoplanets might harbor life or have moons that might be abodes for life? By the end of the twenty-first century, we likely will know.

5

Are Angels Extraterrestrials?

> *And to Allah doth obeisance all that is in the heavens and on earth, whether moving (living) creatures or the angels:*
>
> Qu'ran (16:49)

With the anticipation of having knowledge in hand fairly soon about how many planets in the Milky Way are hospitable for life and perhaps even whether life itself exists beyond the Earth, the remaining chapters examine the theological positions of a number of religions regarding the possibility that extraterrestrial life exists. Before looking in detail at the relevant histories, scriptural texts and theological approaches of these many religions, however, we first need to discuss angels. Specifically, we need to consider whether angels are extraterrestrials.

In most religions angels appear in some form as beings that are able to act as intermediaries between humans and deities. Usually, angels are understood as beings who can travel in between and act in both the spiritual and material worlds. In virtually all cases, angels exist as *supernatural* beings. Angels are essences that are unbound by the natural laws of physics (e.g., gravity) and biology (e.g., eating, breathing, dying) that constrain living things. As such, we can distinguish between living beings that are constrained by the principles that govern material objects in the universe and non-living beings that are unconstrained by those same principles. Our interest in knowing whether extraterrestrial life exists has nothing to do with angels; rather, we are curious to know whether sentient beings exist who, like us, recognize their own mortality as living beings and who may understand the possibility that they might have both physical and spiritual qualities.

Angels are almost always described as purely spiritual beings. They may appear in human form, but they do not have physical bodies made of the same physical essences that compose living things. Moses Maimonides, who built intellectual connections between Aristotelian scholarship and medieval Jewish thought, identified angels as *incorporeal*

D.A. Weintraub, *Religions and Extraterrestrial Life: How Will We Deal With It?*,
Springer Praxis Books, DOI 10.1007/978-3-319-05056-0_5,
© Springer International Publishing Switzerland 2014

beings, as pure spirit, as intelligences (to use Aristotle's descriptive term) without physical bodies. Angels have no material form, no substance.

Though they are not of this world, in Old Testament/Hebrew Bible stories angels often appear and are able to exert a physical presence in this world, despite being incorporeal. Aristotle's intelligences, which were freely incorporated into Christian cosmology, kept the celestial spheres in motion. As told in Genesis, an angel speaks to Abraham and tells him not to kill his son Isaac. Years later, another angel wrestles with Isaac's son Jacob through a night; Jacob may or may not have engaged in a physical wrestling match with a spiritual being, but at dawn he discovers his hip has been badly injured.

In Roman Catholicism, angels are pure spirits created by God who are residents of heaven. Angels who appear in Christian scriptural stories often carry messages from God in heaven to humankind on Earth, but angels do not live on Earth. They can only sojourn in the physical world, and in doing so they leave behind no physical evidence—no footprints, no broken twigs, no bread crumbs—of their visitations. According to these stories, angels can, however, influence events on Earth. As written in the New Testament Gospel of Luke, the angel Gabriel informs Mary that she had found favor with the Lord and would give birth to the Son of God; later, another angel brought news about the birth of a Savior to shepherds tending their flock at night; in the Gospel of Matthew, an angel spoke to Joseph in a dream and urged him to take his wife Mary and the baby Jesus and flee to Egypt.

For both Mormons and Muslims, angels are God's invisible, spiritual messengers. In fact, for both of these religions, an angel helped give birth to the religion itself. According to Islamic tradition the archangel Gabriel spoke with and revealed God's words to Muhammad, thereby inspiring the words that would become the Qu'ran and giving birth to the religion of Islam. Twelve hundred years later, according to Mormon belief, the angel Moroni visited Joseph Smith and continued to visit him over a period of several years until Smith published the Book of Mormon in 1830. Islamic scholars very clearly dismiss the idea that angels are extraterrestrials, as angels, as understood by Muslims, are not alive; angels are not born, they do not die; they do not eat or breathe or walk or swim. Similarly, for Hindus angels are not of the physical world; they occupy a mental rather than a physical plane of existence.

Exceptions to this nearly universal view that angels are something other than physically embodied extraterrestrial beings do exist. Some very conservative Orthodox Christians have taken the position that extraterrestrials have visited Earth and that those visitors are fallen angels, now recognizable as demons trying to lure humanity into the grip of the Antichrist. Some creationist Christians also identify extraterrestrials with angels, though according to these believers these particular angels will not be found on other worlds and will not be detectable by scientists with telescopes.

Whether angels do or do not exist in a particular belief system, the broad consensus across virtually all modern religions and through all of human history is that angels are not extraterrestrials. Extraterrestrials, if they exist, are physical beings. They have a material, tangible essence. They are made of the same basic constituents of the universe as are people, butterflies and dandelions, whether we label those constituents as earth, air, fire and water or as protons, neutrons and electrons, whereas angels are not made of these or any other substances. Finally, if extraterrestrials exist, they should be detectable by humans through their physical interactions with other parts of the material universe.

Part II

Major Religions of the World and Extraterrestrial Life

6

Judaism

*There are Jews—and there are Jews. The Bulbas belong in
the second group.*

William Tenn

Phillip Klass, writing under his pseudonym William Tenn, captures the essence of the Judaic
response to extraterrestrial life in his science fiction story 'On Venus, Have We Got a Rabbi!'[1]
As told by Klass, The Venusian Jewish community has the privilege of hosting the First
Interstellar Neozionist Conference. The Venusian family of Milchik, the TV repairman, has
the honor of hosting three conference delegates from the fourth planet of the star Rigel.
A problem emerges when these delegates from Rigel IV, the Bulbas, turn out not to look
much like Milchik or any other Jews at the conference. Instead, they look like brown, wrin-
kled pillows with gray spots and short tentacles. The conference starts badly. In fact, the
conference cannot start because the Committee on Accreditation refuses to seat the Bulbas.
Even though the Bulbas have appropriate credentials, the Committee on Accreditation concludes that they cannot be Jewish because they do not look Jewish. In particular, they are not
human. The conference is in chaos, unable to seat the Bulbas but also unable to open because
of the credentialing dispute. Rabbi Smallman, from Venus, is appointed to the High
Rabbinical Court that will decide what should be done. The Rabbis think deeply. What would
happen, they ask, if humans go deep into space, perhaps "to another galaxy even, and we find
all kinds of strange creatures who want to become Jews? Suppose we find a thinking entity
whose body is nothing but waves of energy, do we say, no, you're not entirely acceptable?"
Finally, thanks to the wise counsel of Rabbi Smallman, the High Rabbinical Court makes a
decision: "there are Jews—and there are Jews. The Bulbas belong in the second group."

---

[1] Tenn, W. (1974). (Phillip Klass) On Venus, Have We Got a Rabbi! In J. Dann (Ed.), *Wandering Stars*.
New York: Harper and Row.

D.A. Weintraub, *Religions and Extraterrestrial Life: How Will We Deal With It?*,
Springer Praxis Books, DOI 10.1007/978-3-319-05056-0_6,
© Springer International Publishing Switzerland 2014

One can reasonably conclude from this story that Judaism accepts the possibility of extraterrestrial life and that Judaism might be a viable faith for spiritual beings living in other parts of the universe. One might also conclude that Jews have no reason to be at all concerned about those extraterrestrials. Extraterrestrials are God's business. If God wants to make extraterrestrials, God will make them. If God wants them to be Jewish, they can be Jewish. Of course, 'On Venus, Have We Got a Rabbi!' is a work of science fiction. What we want to understand is whether Jewish scripture, history and tradition inform us that this view makes sense. What specifically, if anything, do the writings and teachings of Judaism say about the universe and about life in the universe?

Judaism, the oldest of the western, monotheistic religions, emerged in the early part of the second millennium B. C. E. According to the Jewish tradition, God first spoke to Abram (later Abraham), the first Jew and the patriarch of the Jewish religion, commanding him to pray only to this one God and not the many presumably false gods to whom Abraham's neighbors prayed. God's promise to Abraham was that God would make of him a great nation and bless his descendants. The story of the Jewish people, the descendants of Abraham, of his son Isaac and of Isaac's son Jacob, as remembered and told in the ancient stories about Joseph and his coat of many colors, of the long period of slavery in Egypt, of Moses and the parting of the Red Sea and God's gift to Moses of the tablets on which the Ten Commandments are inscribed, are part of the heritage of all the Western monotheistic religions.

Judaism includes a number of sacred texts known as the Written Law. These texts include the Torah (the books of Genesis, Exodus, Leviticus, Numbers and Deuteronomy), which dates in written form from the fourth to sixth centuries, B. C. E., Prophets (the books of Joshua, Judges, I Samuel, II Samuel, I Kings, II Kings, Isaiah, Jeremiah, Ezekiel, Hosea, Joel, Amos, Obadiah, Jonah, Micah, Nahum, Habukkuk, Zephaniah, Haggai, Zechariah and Malachi) and Writings (the books of Psalms, Proverbs, Job, Song of Songs, Ruth, Lamentations, Ecclesiastes, Esther, Daniel, Ezra and Nehemiah, I Chronicles and II Chronicles).[2]

The Torah includes hundreds of commandments, but those instructions must be interpreted in order for them to be understood. What do the words "You shall not bear false witness against thy neighbor" mean? For more than three millenia, Jewish scholars have attempted to clarify the meaning of the words in the Torah and to explain how the commandments in the Torah are to be carried out.

The Oral Law consists of an enormous set of those commentaries, or rabbinic interpretations. The Oral Law consists of two parts, the Mishna and the Talmud. The Mishna, a legal compilation with some commentaries by learned rabbis on the many laws that appear in the Torah, was first put into written form in about 200 C. E. Over the three centuries that followed, a series of rabbinical commentaries on the Mishna were collected into the set of books known as the Talmud.[3] Orthodox Jews believe that the Oral Law is as sacred as the Written Law, that on Mount Sinai God gave Moses the Oral Law simultaneously with the Torah and that the Oral Law was passed down from generation to

[2] http://www.jewishvirtuallibrary.org/jsource/Judaism/Tanakh.html

[3] http://www.jewishvirtuallibrary.org/jsource/Judaism/Oral_Law.html,   http://www.jewishvirtuallibrary.org/jsource/Talmud/talmudtoc.html, http://www.jewishvirtuallibrary.org/jsource/Bible/jpstoc.html

generation until, thousands of years later, rabbis finally compiled it into written form. Conservative and Reform Jews accept the importance of the Oral Law but believe it is an evolving, human construct rather than a God-given set of explanations.

The idea that all the words in the Torah must be interpreted in order to be understood led not only to the compilation of the Oral Law but to a non-traditional set of scriptures within the Jewish tradition called the Kabbalah, which is used by some but not all Jewish denominations. The writings in the Kabbalah are mystical and concern the nature of the universe and the relationship between the created and the creator. The Zohar is one of the earliest and most important pieces of Kabbalistic writings. It first appeared in the thirteenth century in Spain and was most likely written at that time by Rabbi Moses de Leon.[4] In addition to the Zohar, the Zoharic literature includes the Tikunei Zohar and the Zohar Chadash.

Very few references in any of these Jewish scriptural writings seem to point to concepts even remotely related to extraterrestrial life. In addition, if we are looking for words in scripture that obviously relate to cosmology, we need go no further than the well-known first line from Genesis, "In the beginning when God created the heavens and the earth." This single line of text does not form a strong foundation for asserting anything about the structure of the universe or of the existence of extraterrestrial life; however, some attempts to derive meaning from less obvious pieces of scripture are informative.

One such piece of text is from the Book of Judges, in the Song of Deborah, in which Deborah and her husband Barak, upon defeating Sisera, sing, "They fought from heaven, the stars in their courses fought against Sisera." A few lines later come the words "'Curse ye Meroz', said the angel of HaShem, 'Curse ye bitterly the inhabitants thereof, because they came not to the help of HaShem, to the help of HaShem against the mighty.'"[5] Note that HaShem is not a place. HaShem, a transliteration of the Hebrew equivalent, means "the Name [of God]," and very strict observants use this term to avoid mentioning God in any form. This line of text, then, reveals that an angel of God cursed the inhabitants of Meroz.

Some scholars suggest that Meroz was the name of a city or town destroyed in battle and never rebuilt. The Talmud, in the book Moed Katan, offers two interpretations of Meroz: "Some say that Meroz was [the name of] a great personage; others say that it was [the name of] a star."[6] If Meroz is the name of a star, then Meroz is a star system with extraterrestrial inhabitants. In addition, these extraterrestrials are intelligent and perhaps even have free will, since they chose not to come to the aid of HaShem when they had a choice.

Just 8 months before his death, Rabbi Hasdai Crescas (1340–1410 C. E.), one of the most influential medieval Spanish Jewish philosophers, finished his philosophical work *Light of the Lord*, in which he offered ideas that are fairly similar to those put forward by Nicolaus Cusanus a few decades later. In his anti-Aristotelian, anti-Maimonidean tract, Crescas presents what is likely one of the first Talmudic commentaries on extraterrestrial life. He writes that space is infinite and that infinite space contains a potentially infinite

---

[4] https://www2.kabbalah.com/k/index.php/p = zohar/zohar&vol = 1

[5] Judges 5:20 and 23; *Jewish Publication Society Bible* (1917).

[6] http://www.halakhah.com/

number of worlds. In such a universe, he concludes, nothing in physics and nothing in scriptural or Talmudic writings can deny the existence of extraterrestrial life.[7]

Crescas then recites a Talmudic response to the question "What does God do at night?" by saying "He flies through eighteen thousand worlds on his chariot." These words are a response to a piece of text from the Talmud, in the book Avoda Zara, "He rides a light cherub, and floats in eighteen thousand worlds."[8] The reference to 18,000 worlds emerges, itself, from Jewish mystical tradition in which numerological values are assigned to each Hebrew letter, and the sum of the values of the two letters that spell the word 'life' equals 18.

In his kabbalistic book from the late eighteenth century, Rabbi Pinchas Eliyahu Horowitz of Vilna (1765–1821 C. E.), suggests that intelligent life exists on these 18,000 worlds, though these life forms may not be similar to terrestrial life forms. So why would God ride a chariot through 18,000 worlds, he asks? Because these 18,000 worlds are inhabited by His creations and He chooses to periodically visit them.

According to the Tikunei Zohar, the stars are without number, each star is a world and each righteous man (a tzaddik), of which there are 18,000, will rule over his own world. These 18,000 worlds that God visits daily would be the ones ruled over by each of the 18,000 righteous men. In addition to these worlds, countless other worlds (the stars without number) may exist that are ruled over by the less righteous. The Zohar offers no insight for knowing whether God chooses to visit these other worlds on a daily basis. It also does not make clear how many, if any, of these countless worlds are physical and how many are spiritual or whether that even matters. The Zohar does say that God will give the righteous men, the tzaddikim, wings to escape the Earth and fly through the universe. According to these teachings, the Earth and humanity would have been, initially, the center of God's attention. Over human history the most righteous among the members of Jewish community would have been given the power and privilege of populating these 18,000 other worlds that are scattered throughout the universe.[9]

Another section of scripture that is often cited as having some relevance to Jewish concepts of extraterrestrial life, from the book Berakoth in the Talmud, is the following:

> The Holy One, blessed be He, answered her: My daughter, twelve constellations have I created in the firmament, and for each constellation I have created thirty hosts, and for each host I have created thirty legions, and for each legion I have created thirty cohorts, and for each cohort I have created thirty maniples, and for each maniple I have created thirty camps, and to each camp I have attached three hundred and sixty-five thousands of myriads of stars, corresponding to the days of the solar year, and all of them I have created only for thy sake.[10]

---

[7] Kaplan, A. Retrieved from http://www.askmoses.com/en/article/546,2142111/Extraterrestrial-life.html

[8] http://www.halakhah.com/zarah/zarah_3.html#PARTb

[9] Kaplan, A. *Extraterrestrial life*. Retrieved from http://torahbytes.blogspot.com/2011/09/on-extraterrestrial-life.html

[10] Berakoth 32b; http://www.halakhah.com/berakoth/index.html

Herein, "host" is an archaic term for an army, so this piece of text notes the existence of 30 "armies," or large groups of stars. To continue the military analogy, in Roman history, a legion was a body of the infantry comprised of 3,000–6,000 soldiers. Each legion normally consisted of ten cohorts, each of 300–600 men. Each cohort included three maniples of up to 200 men. If we do the math, we find that these 30 "armies" of stars are composed of hundreds to millions of trillions (10^{14} to 10^{18}) of individual stars. Talmudic scholars have often interpreted the passage "all of them I have created only for thy sake" to mean that while these many other worlds are inhabited, no creatures on any other world have free will.

Rabbi Crescas' student, the Spanish rabbi Yosel Albo (1380–1444 C. E.), asserted that because no other creatures could have free will, there would be no reason for them to even exist; therefore, they do not exist. According to Roger Price, writing on the website *Judaism and Science* (www.judaismandscience.com), four centuries later Rabbi Pinchas Eliyahu Horowitz found the middle ground. In his Torah-based encyclopedia, entitled Sefer HaBris, "he agreed that extraterrestrial beings would have no free will and no moral responsibility, but thought that they might still exist."[11]

Aryeh Kaplan, an American orthodox rabbi of the middle twentieth century (1934–1983 C. E.), who was trained in both physics and kabbalah, agrees with this assessment, writing, "The basic premise that of all possible species only man has free will, is well supported by the great Kabbalist, Rabbi Moshe Kordevero in his Pardes Rimonim. Using tight logical arguments, he demonstrates that there can be only one set of spiritual worlds. Although God would want to maximize the number of recipients of His good, His very unity precludes the existence of more than one such set. Since this set of worlds deals specifically with God's providence toward man because of his free will, this also precludes the existence of another species sharing this quality."[12] According to this interpretation, humankind on Earth is the most special created species in the universe, as we are the only ones with free will. Such a conclusion is in conflict with the cosmological principle, but religious beliefs do not need to conform to the cosmological principle.

The Zohar also offers evidence that some take as authoritative for the existence of extraterrestrial life. In one place, the Zohar teaches that there are seven earths, each with inhabitants. Since these seven earths are separated by a firmament, they are not likely metaphors for the continents on our own planet (of which, until recently, only six had human occupants).

Rabbi Norman Lamm (b. 1927), who served as head of the rabbinic school at Yeshiva University in New York City from 1976 until 2013, is a modern Jewish scholar who has attempted to address, seriously, the possibility that humankind will soon have to face the reality of extraterrestrial life. In his 1986 book *The Religious Implications of Extraterrestrial Life*, he writes, "No religious position is loyally well served by refusing to consider

---

[11] Price, R. New planets, a God for the cosmos and exotheology. In *Judaism and Science*. Retrieved from http://www.judaismandscience.com/new-planets-a-god-for-the-cosmos-and-exotheology/. Accessed 1 Dec 2012.

[12] Kaplan, A. *Extraterrestrial life*. Retrieved from http://torahbytes.blogspot.com/2011/09/on-extraterrestrial-life.html

annoying theories which may well turn out to be facts. ... they [evidence for extraterrestrial life] may be; and Judaism will then have to confront them ..."[13]

What would knowledge about extraterrestrial life mean for humankind and for Judaism? Would the discovery of what Lamm calls other "bio-spiritual residents" in the universe diminish God's interest in us? Would that knowledge diminish the intrinsic worth of humanity? Maimonides never considered man that important in the larger universe, though all other medieval and most present-day Jewish thinkers have argued that God is primarily concerned with the affairs of humankind. Lamm takes a view more consistent with that offered a millennium ago by Maimonides. He asserts, "Man is deemed valuable by Judaism." Our intrinsic worth would be unchanged by knowing that extraterrestrial beings exist: "God is in no way diminished by our learning that His creation far exceeds what had previously been imagined. ... Man's *non-singularity* does not imply his *insignificance*. ... Judaism, therefore, can very well accept a scientific finding that man is not the only intelligent and bio-spiritual resident in God's world. ... we affirm our faith that God is great enough to be concerned with *all* his creatures, no matter how varied and how far-flung throughout the remotest galaxies of His majestic universe. ... man may not *be* the purpose *of* the universe, yet he may *have* a purpose *in* the universe. ... For the believing Jew, therefore, man can accept a far humbler place in the universe than previously assigned to him without surrendering his intrinsic worth and meaningfulness before God."

Should evidence be found that we are not alone, Lamm suggests that at that time "Judaism will then accept the view of one of its most distinguished exponents, Maimonides, over that of the majority with whom he disagreed." Lamm concludes that "The discovery of fellow intelligent creatures elsewhere in the universe, if indeed they do exist, will deepen and broaden our appreciation of the mysteries of the Creator and His creations. Man will be humble, but not humiliated."

So yes, man retains his intrinsic worth. But what about Judaism? Is Judaism a universal religion or one only for humankind on Earth? Lamm's answer is strong and clear: Judaism is only for those who identify as Jews here on Earth: "Torah was given to man on earth and its concern is limited to terrestrial affairs. ..."

Daniel Matt is a leading modern scholar on Jewish mysticism who served on the faculty of the Graduate Theological Union in Berkeley, California for 20 years. Matt describes a different Jewish approach for understanding the relationship between Jews on Earth and any possible extraterrestrial beings we might discover. In *God & The Big Bang*, published in 2001, Matt explains a view of the universe that emerges out of the Zohar, in which God is everything, in which God is the universe.[14] From this perspective, our ability to discover aspects of the universe heretofore unknown (to us) is identical with learning more about God. Matt writes that, in the words of Rabbi Dov Baer (d. 1772 C. E.), one of the late eighteenth-century founders of Hasidism in the Ukraine, "When you gaze at an object, you bring blessing to it."[15] Matt explains this further: "God hides within each of us, within all of creation and throughout spacetime." He continues, saying "Divinity pervades the

---

[13] Lamm, N. (1986). *Faith and Doubt: Studies in Traditional Jewish Thought* (2nd ed.) New York: KTAV Publishing House.

[14] Matt, D. C. (2001). *God & the Big Bang*. Woodstock, VT: Jewish Lights Publishing.

[15] As quoted in *God & the Big Bang* (p. 78).

universe; its sparks animate every single thing."[16] From this mystical point of view, by discovering extraterrestrial life we would be discovering more of God's creations and thereby more about God. Clearly, the discovery of extraterrestrial life would be welcome to any Jews who follow this line of spiritual thinking.

One overarching theme within Judaism is that there are no limits on the power of the creator. Thus, for Jews to say that no life beyond the Earth could possibly exist would be unacceptable, as such an idea would appear to place shackles on God's creative power. The Judaic approach reduces to this: the universe belongs to God (or is God) and God can do what God wishes to do with the universe. We can conclude that although some of the Jewish canon offers strong suggestions that we are not alone, Judaism draws no firm conclusions about the existence of extraterrestrial life, and Jewish beliefs do not depend in any way on the existence of extraterrestrials. The Jewish answer appears to be thus: maybe somebody is out there or maybe not, maybe they have free will or maybe not; that's up to God.

Even without a definitive answer as to how Jews around the world might feel about the existence or non-existence of extraterrestrial life, we can ask how any answer to that question would impact Judaism. Scriptural stories embraced by Jews include a number of persons and events that are presumed to be historical. The practice of Judaism is often centered on remembering many of these persons and events (the holidays of Purim, Hannukah and Passover, for example); these holidays of remembrance have as much to do with cultural practice as with religious belief, although the two are intimately linked. The existence—or non-existence—of extraterrestrials would not clearly affect Jewish duties or practices and likely would have no effect on beliefs.

But could extraterrestrials be Jewish? What about William Tenn's Bulbas from Rigel IV? Are the sites of the First and Second Temples in Jerusalem and of the siege at Masada important in Jewish history? Yes, for Jews on Earth. Could a living being from another part of the universe know nothing about Adam, Noah, Abraham or Moses, or about the festivals of Purim and Passover, and still be Jewish? No.

Extraterrestrials might discover, through their own history and prophets, ways to worship the same God, but they would not be Jewish. In Jewish practice, Judaism is a way of life for those who identify as Jews that enables them to develop a relationship with God. Jews, however, would never presume to limit God's power. Elsewhere in the universe, God might choose to guide others along a different path into a worshipful relationship with God, and if that relationship has a different label, *mazel tov*.

According to Jewish belief, is the God of Judaism universal? Yes. Is Judaism universal? Not likely. Jews have no idea and perhaps, to many of them, the answer doesn't matter, but the answer does matter to Jeremy Kalmanofsky, a faculty member of The Jewish Theological Seminary and a practicing rabbi in New York City. Kalmanofsky has eloquently written, in his 2013 essay "Cosmic Theology and Earthly Religion," "The cosmos is so vast it silences the mind. ... We live on a nothing-special planet, orbiting a nothing-special star ... Theologically, I must conclude that this all cannot be about us... We still theologize geocentrically. Contemporary Judaism needs a faith befitting our cosmos; a faith that does not narrow the infinite God to the infinitesimal conditions of our times and

---

[16] *God & the Big Bang* (pp. 79, 135).

places." He continues, saying "Religion is the human, social response to transcendence. ...
Normative Judaism provides an excellent, time-tested path for sanctifying our minds, morals, and bodies, refining us as a people, improving the world, correlating our lives to the
infinite God unfolding on the finite earth."[17]

So yes, the God of Judaism is universal, but no, Judaism is not. Judaism is for humans
on Earth.

Michael Ashkenazi, an anthropologist and senior research scholar at the Bonn
International Center for Conversion (The BICC conducts no religious activities; the word
'conversion' in the Center name refers to the conversion of military bases to peaceful uses),
writing in 'Not the Sons of Adam,' points out that Judaism is about the relationship between
Jews and God, a God who is universal and unique: "There is no clear indication in Judaism
that man is the only rational being, nor that man, as man, has a unique and primary relationship to God. Moreover, God's universal primacy and uniqueness are unquestionable.
The possible existence of ETI [Extra-Terrestrial Intelligence] is thus not problematic, as
other intelligent beings would be considered merely as God's creations. The relationship
between Jews and God, on the other hand, is a very specific one, defined in behavioural
terms, and the existence of other life forms, newly discovered scientific realities or pan-
human behavioural changes would not affect that relationship in the slightest."[18]

The following story[19] sums up the Jewish attitude well: During a lecture by a rabbi, an
unstable-acting man in the crowd continually got up to interrupt the rabbi and kept insisting that the rabbi must agree with him that extraterrestrial life exists. The rabbi finally
replied to the meshugener (a crazy or addled person), "Okay! Fine! Extraterrestrial beings
exist! But what does that have to do with the fact that you're not learning any Torah?"

---

[17] Kalmanofsky, J. (2013). Cosmic theology and earthly religion. In E. J. Cosgrove (Ed.), *Jewish Theology in Our Time* (pp. 23–30). Woodstock, VT: Jewish Lights Publishing.

[18] Ashkenazi, M. (1992, November). Not the sons of Adam. *Space Policy* (p. 343).

[19] As told to the author by Barbara Bensoussan.

7

Setting the Stage for Modern Christianity

> *To believe that God created a plurality of worlds, at least as*
> *numerous as what we call stars, renders the Christian*
> *system of faith at once little and ridiculous.*
>
> Thomas Paine (Paine, T. (1796). *The Writings*
> *of Thomas Paine: The Age of Reason—Part I and II*, vol. IV.
> Collected and edited by Moncure Daniel Conway.
> Project Gutenberg ebook; 2012.)

Christianity emerged two millennia ago, as the disciples of Jesus of Nazareth developed and codified a set of beliefs, built around Jesus's life, teachings, death and reported resurrection. Over a period of several centuries, early Christian leaders determined which books and other writings would be included in what would eventually be known as the Roman Catholic Bible. By the end of the fourth century, C. E., the contents of the Roman Catholic Old and New Testaments had been canonized.

According to Roman Catholic belief, the books in the Old and New Testaments are divinely inspired and the inerrant Word of God, as communicated through prophets. When interpreted by religious authorities, the truths found in these words are considered sacred. For Roman Catholics, the popes and Councils of the Church are the only authoritative interpreters of scripture.

Roman Catholics believe that humans are intrinsically good; however, in the Garden of Eden Adam sinned when he defied God's instructions by eating fruit from the tree of knowledge of good and evil. According to Roman Catholic belief, this defiant act, known as original sin, causes all humans to be born corrupted by sin. The issue as to whether Adam's act of defiance took place in an actual garden on Earth several thousand years before today is an important one for Christianity in general and for understanding how Roman Catholics understand extraterrestrial life.

D.A. Weintraub, *Religions and Extraterrestrial Life: How Will We Deal With It?*,
Springer Praxis Books, DOI 10.1007/978-3-319-05056-0_7,
© Springer International Publishing Switzerland 2014

Roman Catholics believe that one's lifetime on Earth is an opportunity, that individual's only opportunity, to earn God's gift of salvation by cleansing one's soul of sin. According to Roman Catholic belief, the souls of Catholics who are saved will spend eternity in heaven while those who do not achieve salvation will suffer for eternity in hell. In Roman Catholic belief, heaven is a physical location, the dwelling place of God, Christ, the saints and the angels, whose location is not identified other than that is located beyond the Earth. Hell is also a physical location, the one to which the damned descend, a place isolated from the light of heaven. Though the Roman Catholic Church has never decided exactly where hell is, through most of the last 2,000 years, most theologians have generally accepted the opinion that hell is within the Earth. Thus, in an odd way, humans on earth, in comparison to extraterrestrials, live in a very privileged place with respect to the location of hell (though not with respect to the unknown location of heaven), unless every world has its own hell. This is not a theological issue that has received any attention.

Roman Catholics also believe that God's only begotten son entered the human world about 2,000 years ago in the form of the human being identified as Jesus of Nazareth, and that about 30 years after his incarnation Jesus sacrificed himself on the cross in order to redeem all of humanity from original sin. Notably, Jesus did not perish on the cross to redeem any single person for the individual sins committed by that person during his or her lifetime, nor did he die "for the dolphins or the gorillas, and certainly not for the proverbial little green men."[1] Rather, Jesus redeemed every believing Catholic for the sinful condition into which Catholics believe all persons are born, as a result of the original sin of Adam.

The commission of original sin and the crucifixion of Jesus are intimately and inextricably linked in Roman Catholic theology. In the context of extraterrestrial life, a reasonable question to ask is whether those two acts define Roman Catholicism for any and all living beings in the universe or only define Roman Catholicism for terrestrial beings.

While still a resident of the original paradise, Adam violated God's instructions not to eat from the tree of the knowledge of good and evil in the Garden of Eden. As told in the Book of Genesis, the act of eating fruit from the tree of knowledge allowed evil and death to become part of the human experience. According to the Apostle Paul, this act was the original sin: "Wherefore as by one man sin entered into this world and by sin death: and so death passed upon all men, in whom all have sinned."[2] Thus, according to Roman Catholic belief, all humans are sinful from birth and can only be redeemed from original sin by accepting Jesus as Savior and redeemer.

Roman Catholic scripture does not identify any other forms of life on Earth as sinful or in need of or as capable of redemption; original sin does not apply to whales or mice or trees. These living beings on Earth presumably are neither sinful nor in need of redemption. As Adam and Eve were humans who sinned in the Garden of Eden on Earth, one might presume that non-human extraterrestrials would not be contaminated by original sin and would not be in need of redemption by Jesus. If this is the correct way to understand the issue of original sin, then humans could be the only sinful creatures in the universe and the only species in need of redemption. Since salvation, as understood in Catholicism, is

---

[1] Davies, P. (2003, September) E.T. and God. *The Atlantic Monthly*.
[2] Romans 5:12.

obtained through the redemptive actions of Jesus, this understanding of original sin could mean that no other creatures in the universe need to be saved and therefore that no other creatures can be saved and no extraterrestrials can achieve eternal life. Alternatively, perhaps all other species throughout the universe are saintly, without sin and without need for redemption, and perhaps they achieve eternal life without salvation. If no other beings in the universe are corrupted by evil and therefore have no need for redemption, can those creatures be Catholic? Finally, perhaps other species, or at least some other species, are sinful and in need of salvation, either because they are somehow affected by the original sin of Adam or because original sin occurs in their own worlds through a different mechanism.

If original sin is built into the very fabric of the universe in such a way that all beings who might exist anywhere in the universe could be affected by that evil, then original sin could be a universal issue and the events that took place in the Garden of Eden would not create any theological problems with regard to extraterrestrial beings; however, if original sin is universal, then the need for redemption from those sins and salvation is also universal. If so, what is God's salvation plan for extraterrestrials? In the words of the very influential French Dominican priest and theologian Cardinal Yves Congar (1904–1995 C. E.), "Christianity is one plan for salvation that has been revealed by the Father."[3] Congar's words imply that other plans—ones that do not necessarily involve incarnations, let alone Jesus, ones that would not be labeled as Christianity—may exist for other species in need.

Perhaps the first pre-Copernican-era Roman Catholic theologian to expound positively on the possibility that extraterrestrial life could exist and to address the issues of original sin and the crucifixion of Christ in this context was the highly regarded Franciscan scholar Guillaume de Vaurouillon (c. 1392–1463 C. E.; also identified as William or Willem van Vorilong). In contradiction to the by-then widely accepted views of Thomas Aquinas, Vaurouillon made clear his learned opinion that other worlds could exist: "He could create an infinity of worlds better than this one."[4]

He further wrote that life is likely to exist on those worlds: "Infinite worlds, more perfect than this one, lie hid in the mind of God ... it would be possible for the species of this world to be distinguished from that of the other world." But in Vaurouillon's view these other beings did not suffer from original sin: "If it be inquired whether men exist on that world, and whether they have sinned as Adam sinned, I answer no, for they would not exist in sin and did not spring from Adam."

Though Vaurouillon does not make clear why he believes these extraterrestrials would be in need of redemption since he has asserted that they do not suffer from the sins of Adam, the extraterrestrials, in his view, would nevertheless be redeemed by the one-time death and one-time resurrection of Christ that occurred on Earth, though he left the mechanics of how this could take place unexplained: "As to the question whether Christ by dying on this earth could redeem the inhabitants of another world, I answer that he is able to do this even if the worlds were infinite, but it would not be fitting for Him to go into another world that he must die again."

---

[3] As quoted in O'Meara, T. F. (2012). *Vast Universe* (p. 91). Collegeville, MN: Liturgical Press.

[4] As quoted in McColley, G., & Miller, H. W. (1937). Saint Bonaventure, Francis Mayron, William Vorilong, and the doctrine of a plurality of worlds. *Speculum, 12*(3), 386.

The Protestant Reformation, triggered by the actions of Martin Luther in 1517, split the Church of Rome (i.e., Roman Catholicism) into warring factions; however, the debate concerning original sin, the crucifixion and extraterrestrial life remained a centerpiece of Christian intellectualism, not one that had a home exclusively within either the Catholic or Protestant domain.

Philip Melanchthon (1497–1560 C. E.), a colleague of Martin Luther and an intellectual leader of the Lutheran Reformation, spoke out forcefully in opposition to the idea of a pluralistic, heliocentric universe, in large part because "The Son of God is one: our master Jesus Christ, coming forth in this world, died and was resurrected only once. Nor did he manifest himself elsewhere, nor has he died or been resurrected elsewhere. We should not imagine many worlds because we ought not imagine that Christ died and was risen often; nor should it be thought that in any other world without the knowledge of the Son of God that men would be restored to eternal life."[5]

For Melanchthon, as for Vaurouillon, the only path to salvation is through Jesus Christ, and they agree that the resurrection of the Son of God took place on the Earth and only on the Earth. Their understandings of these ideas differ dramatically, however; whereas Vaurouillon allows that the saving grace of Jesus will somehow reach all extraterrestrials in need, Melanchthon believes that extraterrestrials would never be able to learn about Jesus. Melanchthon takes his logic one step further: since without the saving grace of Jesus gained through the resurrection salvation cannot take place, extraterrestrial life cannot exist.

Melanchthon's words are interesting because he does not say explicitly that other worlds and multiple crucifixions are impossible; instead, he posits that we should not even think about the possibility of other worlds. Perhaps Melanchthon thought that if we even dare to imagine the crucifixion happening multiple times in multiple places, our mere thoughts about such possibilities might make them happen. One way to understand Melanchthon's logic is to recognize that, from his point of view, if life exists beyond the Earth, those extraterrestrials could not possibly enjoy the benefits of a Christian life, death or afterlife. Since Christianity, understood in this way, is not possible anywhere but on Earth and since life should not exist without Christianity, life on Earth should be unique.

The Dominican Friar Tommaso Campanella (1568–1634 C. E.) wrote *Apologia pro Galileo* (The Defense of Galileo) in 1616, while imprisoned in Naples (for 27 years) for various heresies and for conspiring against Spanish rule. In this treatise, first published in 1622, he supports the doctrine of the plurality of worlds and rejects the idea that original sin applies to any created beings in the universe other than earthlings or that extraterrestrials require redemption: "If the inhabitants which may be in other stars are men, they did not originate from Adam and are not infected by his sin. Nor do these inhabitants need redemption, unless they have committed some other sin."[6]

The Roman Catholic Church had been down this road before. Lactantius (240–320 C. E.), advisor to Emperor Constantine, had denied the existence of the Antipodes (a

---

[5] *Initia doctrinae physicae*. Corpus Reformatorum, *13*(1), 221, as quoted in O'Meara, T. F. (1996). Christian theology and extraterrestrial intelligent life. *Theological Studies, 60*, 6.

[6] Campanella, T. (1937). *The Defense of Galileo, Mathematician of Florence* (McColley, G., Trans.) In *Smith College Studies in History* (vol. XXII, Nos. 3–4, p. 66).

location on the bottom of a spherical Earth, assuming Europe would be on the top), let alone the existence of humans living in such a location, saying "Can anyone be so foolish as to believe that there are men whose feet are higher than their heads, or places where things may be hanging downwards, trees growing backwards, or rain falling upwards?"[7] A few decades later, his assertion was strongly supported by Saint Augustine, who cited the nineteenth Psalm and the Apostle Paul's Epistle to the Romans to declare that inhabited antipodes would be contrary to scripture, as they could not be inhabited by descendants of Adam. Other Church leaders added to this conjecture by arguing that if any humans lived at the antipodes, then Christ would "of necessity be crucified both here and there."[8] In their view the idea of multiple crucifixions was absurd. Three centuries later, in 748 C. E., Pope Zachary (679–752 C. E.; pope from 741 to 752 C. E.) declared the Antipodean Heresy, asserting that "if it shall be clearly established that he professes belief in another world and other people existing beneath the earth, or in [another] sun and moon there, thou art to hold a council, and deprive him of his sacerdotal rank, and expel him from the church."[9] By the time of Galileo and Campanella, however, Christopher Columbus had sailed to the new world (1492 C. E.), Ferdinand Magellan's expedition had circumnavigated the globe (1519–1522 C. E.) and humans had been found living, effectively, in the antipodes. As a result, the Antipodean Heresy faded away. With the discovery of heretofore unknown human populations, both Roman Catholic and Protestant leaders drew the same conclusion: the salvation of those people who previously had no knowledge of Christ necessitated proselytizing and converting them in order to offer them the gift of eternal life. The questions as to whether these new world peoples were descended from Adam and had suffered original sin and how such things might have happened were left unanswered. Thus, the early lesson of Christians discovering sentient beings living in 'new worlds' was to assert that the newly discovered life forms had no prior knowledge of the crucifixion and resurrection of Christ, but nevertheless did need to be converted to Christianity in order for them to be granted eternal life in heaven.

Gottfried Wilhelm Leibniz (1646–1716 C. E.), born in Protestant Germany but perhaps best characterized as a mystical Christian deist rather than as a Lutheran or Catholic, believed in extraterrestrial life but doubted that original sin and Christian redemption applied to extraterrestrial beings. In his *Nouveaux essais sur l'entendement humain* (New Essays Concerning Human Understanding) (written in 1704 but not published until 1765 C. E.), he wrote,

If someone else came from the moon in some extraordinary machine, like Gonzales,[10] and told us credible things about his homeland, we would take him to be a lunarian;

---

[7] Lactantius, quoted in *The Discoverers* (p. 107), by Boorstin D. J. (1983). New York: Random House.

[8] Campanella, T. (1937). *The Defense of Galileo, Mathematician of Florence* (G. McColley, Trans.) In *Smith College Studies in History* (vol. XXII, Nos. 3–4, p. 15).

[9] *Monumenta Germaniae Historica, Epistolae Selectae*, 1, 80 (pp. 178–179) (M. L. W. Laistner, Trans.) In *Thought and Letters in Western Europe* (pp. 184–185).

[10] A reference to the fictional hero of *The Man in the Moone, or a Discourse of a Voyage thither by Domingo Gonsales* (London, 1638), a novel written by the English bishop and science fiction writer Francis Godwin.

and yet we might grant him the rights of a native and of a citizen in our society, as well as the title *man*, despite the fact that he was a stranger to our globe; but if he asked to be baptized, and to be regarded as a convert to our faith, I believe that we would see great disputes arising among theologians. And if relations were opened up between ourselves and these planetary men—whom M. Huygens[11] says are not much different from men here—the problem would warrant calling an Ecumenical Council to determine whether we should undertake the propagation of the faith in regions beyond our globe. No doubt some would maintain that rational animals from those lands, not being descended from Adam, do not partake of redemption by Jesus Christ; but perhaps others would say that we do not know enough about all the places that Adam was ever in, or about what has become of his descendants—for there have even been theologians who thought that Paradise was located on the moon. Perhaps there would be a majority decision in favour of the safest course, which would be to baptize these suspect humans conditionally on their being baptizable. But I doubt they would ever be found acceptable as priests of the Roman Church, because until there was some revelation their consecrations would always be suspect.[12]

Leibniz's position appears to be one that limits Christianity to Earth. In his view, Christianity is a terrestrial religion, only. Jesus died on the cross only for humanity. Those who agree with Leibniz might also imagine that extraterrestrials worship the same God, but that they need not be and probably cannot be turned into Christians.

The Revolutionary War era polemicist Thomas Paine (1737–1809 C. E.), writing in *Age of Reason* (1793), articulated clearly and directly the most forceful arguments against Christianity as a universal religion under the assumption that extraterrestrial life exists:

Though it is not a direct article of the Christian system that this world that we inhabit is the whole of the habitable creation, yet it is so worked up therewith from what is called the Mosaic account of the Creation, the story of Eve and the apple, and the counterpart of that story, the death of the Son of God, that to believe otherwise—that is, to believe that God created a plurality of worlds, at least as numerous as what we call stars—renders the Christian system of faith at once little and ridiculous and scatters it in the mind like feathers in the air. The two beliefs cannot be held together in the same mind, and he who thinks that he believes both has thought but little of either.[13]

From whence then could arise the solitary and strange conceit that the Almighty, who had millions of worlds equally dependent on his protection, should quit the care of all the rest, and come to die in our world, because, they say, one man and one woman had eaten an apple. And, on the other hand, are we to suppose that every world in the boundless creation had an Eve, an apple, a serpent, and a Redeemer? In this

---

[11] Dutch astronomer Christian Huygens (1629–1695 C. E.), who discovered the rings of Saturn and Saturn's largest moon Titan.

[12] Leibniz, G. W. (1949). *New Essays Concerning Human Understanding* (p. 342). (A. G. Langley, Trans.) La Salle, IL: The Open Court Publishing Company.

[13] Paine, T. (1880). *The Age of Reason* (p. 38). London: Freethought Publishing Company.

case, the person who is irreverently called the Son of God, and sometimes God himself, would have nothing else to do than to travel from world to world in an endless succession of deaths, with scarcely a momentary interval of life.[14]

But such is the strange construction of the Christian system of faith, that every evidence the heavens affords to man, either directly contradicts or renders it absurd.[15]

Whether one is a Roman Catholic, Lutheran, Methodist, Presbyterian, Calvinist, or an adherent to the doctrinal beliefs of any other Christian denomination, in regard to extraterrestrial life and within the framework of a very "simplistic Christology,"[16] Paine has framed the question exquisitely well. A believing Christian has to choose between these two alternatives:

1. Christianity is not a universal religion. The sacrifice of Christ on the cross was a redemptive event that took place on the Earth and that occurred in response to an earlier event that also took place on Earth, in the Garden of Eden. Embracing this redemptive act of Christ is an absolute requirement to be a Christian. Jesus' death was a singular act meant to redeem humans and only humans.
2. Christianity is a universal religion. If other sinful and sentient species exist, wherever they are in the universe they somehow will learn the redemptive power of the singular event that occurred on Earth, that of the crucified and risen Christ.

If Christianity is a religion limited to life on Earth and if Christianity is the only spiritual path in the universe to God, the logical conclusion from this point of view is that the only God-loving beings in the universe who are capable of a spiritual life must live on Earth. After all, a just, loving and all-powerful God surely would not create rational creatures elsewhere in the universe only to abandon them by offering them no hope of eternal life. This view is self-contradictory, however, because an all-powerful God should be able to offer extraterrestrials alternate paths to eternal life and some modern theologians have been seeking to identify those alternate paths.

On the other hand, if the opossums and wasps and sharks on Earth do not have the opportunity that humans have for eternal life, perhaps life without the possibility of redemption for extraterrestrials is consistent with this view of Christianity. In this case, extraterrestrials could exist and would be like non-human animals on earth—without sin and without need of redemption. Perhaps humans are the worst, most evil, depraved and sinful creatures in all of God's creation; hence, humans are the only species in the universe in need of Christ's redemptive act. This view of Christianity does not depend on the existence or non-existence of extraterrestrial life and might be unaffected by any discoveries made by astronomers.

If one chooses to believe in a universe in which Christianity, including the necessity of multiple incarnations and crucifixions, is universal, one must then embrace a view of Christianity that Paine described as "an endless succession of death."

---

[14] Ibid, p. 44.

[15] Ibid, p. 45.

[16] O'Meara, T. (2012). *Vast Universe* (p. 82). Collegeville MN: Liturgical Press.

In fact, no modern theologians would argue for multiple crucifixions, though quite a few find the concept of multiple incarnations acceptable. A few modern thinkers have tried to address this issue, though most of them have devoted very little thought to solving the theological problems. According to the liberal Anglican priest-theologian and Dutch biochemist Sjoerd L. Bonting (b. 1925),[17] the list of contemporary theologians and theologian/scientists who are silent on the matter of extraterrestrial life—he specifically identifies Karl Barth (Swiss Protestant Reformed), Emil Brunner (Swiss Protestant Reformed), Hans Küng (Swiss Catholic), John Macquarrie (British Episcopalian), Wolfhart Pannenberg (German Lutheran), Jürgen Moltmann (German Protestant Reformed), Edward Schillebeeckx (Belgian Roman Catholic), Keith Ward (British Anglican), Arthur Peacocke (British Anglican), John Polkinghorne (British Anglican) in this group—far exceeds the few who have had at least a little to say. We would be remiss, though, in not responding to Bonting by saying that the number of recent and contemporary theologians who have or who are addressing the issue of extraterrestrial life is growing and that some of the theologians in his list have been soft spoken on the issue rather than completely silent.

How different Christian denominations understand both the crucifixion of Jesus and the religious meaning of the sins committed by Adam and Eve in the Garden of Eden helps distinguish them from one another. These understandings also reveal a great deal about how each Christian denomination will react to the discovery of extraterrestrial living beings.

---

[17] Bonting, S. L. (2003, September). Theological implications of possible extraterrestrial life. *Zygon 38*(3).

8

Roman Catholicism

> *Is the Bible such an inspired book that the only salvation*
> *history possible is the one that it records, and is Jesus so*
> *central a figure that all intelligent species away from earth*
> *must include knowledge of him?*
>
> Thomas F. O'Meara (O'Meara, T. F. (1999).
> Christian theology and extraterrestrial intelligent life.
> *Theological Studies, 60*, 3.)

In her science fiction novel *The Sparrow*,[1] Mary Doria Russell takes on a religious idea that has been deeply entrenched within the monotheistic religions: special election. Are human beings, or is even a religious subgroup of humanity, specially chosen by God? Are Jews the people of the covenant? Are Christians, or certain Christians, specially selected by God to be redeemed? Have Muslims received a special promise from God? Are Mormons the chosen people? If the members of a religious group believe they have been specially blessed by their God, whom they believe is the God of and the creator of the entire universe, does their covenant with God include an obligation and burden to inform others as to how to become one of the elect? If proselytizing is part of the mission for this group of believers, how would that activity be carried out across interstellar and intergalactic distances?

In *The Sparrow*, astronomers on Earth in the year 2019 intercept and decipher messages sent in the form of music and poetry from the planet Rakhat in the Alpha Centauri system. Soon thereafter, several Jesuit priests lead a mission to Rakhat in order to find and meet the beings who are broadcasting these messages. Though the Jesuits priests who travel to Rakhat do not claim to be missionaries, the Earth-bound Jesuit leaders who fund,

---

[1] Russell, M. D. (1996). *The sparrow*. New York: Fawcett Columbine.

D.A. Weintraub, *Religions and Extraterrestrial Life: How Will We Deal With It?*,
Springer Praxis Books, DOI 10.1007/978-3-319-05056-0_8,
© Springer International Publishing Switzerland 2014

staff and monitor the mission clearly see the expedition as something with a purpose much greater than mere scientific and cultural exploration. The hazards the humans must deal with when they attempt to engage with the natives on Rakhat resemble some of the tragic clashes that occurred in our own recent history when European missionaries first encountered the native peoples of North America, South America, Africa and Asia.

If all beings have in common creation by the same God, yet one group of people on Earth believe they have been specially selected by God, via a revelation from God to one of their prophets or a visit to them by the Son of God, will the selected people's understanding of their special relationship with God be affected by knowing extraterrestrial Others are out there? The Roman Catholic intellectual tradition offers a 1,000 years of on-again, off-again theological debate on the subject of the possible uniqueness and centrality of humankind in the universe and of Christianity for humankind—and by extension with Others. The result of this lengthy conversation is that in the modern Roman Catholic view extraterrestrial creatures could and probably do exist. On this point, twentieth- and twenty-first-century Roman Catholic scholars are in broad agreement, though there appear to be some prominent exceptions to the rule. As for the spiritual nature of those beings— are they, like humans, sinful beings in need of redemption and salvation?—the disagreements among leading modern Roman Catholic scholars are great. As for the desire or need for missionary activities, no modern Catholic intellectuals who write about the possibility of extraterrestrial life imagine conversion to be a plausible or even a comprehensible goal of future contact with extraterrestrials. But even without imagining that future Jesuits, or any other Catholic missionaries, would attempt to convert a non-terrestrial group of living beings to Catholicism, the question "How can a particular group of human beings remain God's chosen people once we know that God also created intelligent beings who are not and cannot become part of our religious community and who are perhaps far more intellectually capable and technologically advanced than are we?" remains one worthy of theological inquiry and one well within the broad interests of Roman Catholic scholars.

Roman Catholic leaders have expressed, with increasing levels of certitude over the last two centuries, confidence that extraterrestrial life can exist and that their existence offers no insurmountable obstacles to Catholic beliefs. For example, Theodore Hesburgh (b. 1917), who served as President of the University of Notre Dame from 1952 until 1987, suggested in 1977 that God is likely to have created intelligent extraterrestrial beings: "What reflects God most is intelligence and freedom, not matter. Why suppose that He did not create the most of what reflects Him the best? He certainly made a lot of matter. Why not more intelligence, more free beings, who alone can seek and know him? ... Finding others than ourselves would mean knowing Him better."[2] Though Hesburgh eagerly embraces the idea that extraterrestrials exist, he avoids addressing the complicated doctrinal issues associated with Roman Catholic theology and practice that surround this view.

A few years later, in 1984, Swiss Catholic priest and theologian Hans Küng (b. 1928) wrote similarly, "We must allow for living beings, intelligent—although quite

---

[2] Hesburgh, T. (1977). Forward. In P. Morrison, et al. (Eds.), *The search for extraterrestrial intelligence (SETI)*. NASA, as quoted in O'Meara, T. F. (2012). *Vast universe: Extraterrestrials and Christian revelation* (p. 15). Collegeville, MN: Liturgical Press.

different—living beings, also on other stars of the immense universe."[3] Küng tells us nothing about the spiritual needs of these distant, living beings and gives us no help in understanding whether Christianity, or specifically Roman Catholicism, would play any role in the spiritual lives of these beings. He implicitly and firmly asserts, however, that from a Roman Catholic point of view, such beings could exist without affecting Christian beliefs at all.

The view that extraterrestrial life can exist, even that it is likely to exist, emerged in the serious theological writings of a very few scholars within the Roman Catholic community in the second half of the nineteenth century. The writers who presented this point of view also began to recognize and wrestle with the serious theological issues that might appear to be in conflict with this idea, i.e., with the challenge issued two centuries ago by Thomas Paine.

In 1889, a New Jersey priest and scholar Januarius De Concilio (1836–1898 C. E.), writing in his book *Harmony Between Science and Revelation*, offered a number of arguments in support of the existence of extraterrestrial beings. Specifically, he wrote about creatures "with intelligent substances united to some kind of body," that is, beings that are something other than the pure spiritual beings known to Catholics as angels. One reason he forcefully offers against the argument that such beings cannot exist is that "Revelation never taught, nor obliges any one to hold, such a thing [that man is the center and end of all creation]." From De Concilio's perspective, if Revelation does not preclude the existence of extraterrestrials then they can exist. De Concilio then describes how Abbé Moigno (1804–1884 C. E.), a French Jesuit priest and scientist and founder in 1852 of the scientific journal *Cosmos*, was authorized by the Sacred Congregation of the Index (the committee of Cardinals that determines whether books are opposed to Catholic morals and truth; established in 1571 and merged into the Supreme Sacred Congregation of the Holy Office in 1916) to determine if the plurality of worlds doctrine was opposed to any Catholic doctrine. Abbé Moigno concluded that the plurality of worlds doctrine "did in no way conflict with the doctrines of the Creation, Incarnation, and Redemption as taught by the Catholic Church."

De Concilio then offers his explanations regarding sin, redemption and the role of the sacrifice of Christ on Earth for the salvation of souls throughout the universe. All "inhabitants of the stars, we may easily surmise, were created by God." Such a statement, according to De Concilio, is based directly on the New Testament verses John 1:3 ("All things came into being through him, and without him not one thing came into being") and Colossians 1:16 ("For in him all things in heaven and on earth were created, things visible and invisible, whether thrones or dominions or rulers or powers—all things have been created through him and for him"). De Concilio continues, arguing that if these imagined inhabitants of stars have sinned, which he does not assume to necessarily be the case, "they, like all other created personalities, must come in communion with Christ in order to partake of the sublimation of the whole universe, and in order to be able to reach their everlasting destiny." Just in case the idea of recognizing Christ as a universal savior was not clear, De Concilio repeats this assertion: sinful inhabitants of other worlds "cannot

---

[3] Küng, H. (1984). *Eternal life?* (p. 224) New York: Doubleday, as quoted in Peters, T. (2003). *Science, theology, and ethics* (p. 126). Ashgate science and religion series.

attain their destiny except by being united with Christ." Very clearly, in the view of De Concilio Christianity, that is, "being united with Christ," is the sole path to salvation available to sentient beings throughout the universe.

How could intelligent, spiritual beings on other worlds know about Christ? Was Christ incarnated and resurrected in the flesh of the extraterrestrial Others as well as in the body of Jesus? Or do they somehow need to learn about the life and death of Jesus, about events that happened on the physical Earth 2,000 years ago? Fortunately for those creatures not living on Earth, De Concilio has an answer. "When Christ died and paid the ransom of our redemption, He included them also in that ransom, the value of which was infinite and capable of redeeming innumerable worlds." This answer is a firm 'no' to the question, "Did multiple resurrections happen?" How news or knowledge of Christ's ransom will be transmitted to beings on other worlds is clearly a matter of faith, not of science, but De Concilio asserts that "God, in His infinite goodness, wanted to make the universe an infinite expression of Himself, at least by union. His divine Word, the infinite expression of his grandeur, came to reside in the universe by uniting to Himself the human nature in the body of His own personality, and thus He divinized the whole universe, inasmuch as human nature represented all its existing species." Thus, the Son of God was only incarnate once, as a human, on one world, the Earth. And Jesus was only crucified once, but that single event provides redemption for all beings throughout the universe across all times.[4] De Concilio thus avoids Thomas Paine's trap of multiple crucifixions as God's mechanism for redeeming creatures on other worlds, but if one is looking for a mechanism that does not violate the laws of physics for both space and time for getting information about Christ from Earth to other worlds, that answer cannot be found in the writings of De Concilio. In addition, this answer leaves no doubt that the Earth is a place of unparalleled importance and humanity is a species of unmatched privileged—after all, the crucifixion occurred on Earth—in a universe that includes Others. De Concilio's universe is spiritually, even if not physically, geocentric.

In his book *Vast Universe: Extraterrestrials and Christian Revelation*, published in 2012, Dominican theologian Thomas F. O'Meara, now Professor Emeritus of the University of Notre Dame, identifies Joseph Pohle (1852–1922 C. E.) as another prominent nineteenth-century theologian who willingly confronted the cosmotheological issues associated with the existence of extraterrestrial Others. Pohle is the author of the 12-volume *Dogmatic Theology* and was one of the first members of the faculty of Catholic University of America in Washington, D.C. In his 1884 book *Star-worlds and Their Inhabitants*, Pohle speaks out in favor of the existence of extraterrestrials, largely because of the "full analogy between our earth and many other celestial bodies."[5] For Pohle, the analogy has limits, however, as "no reason compels us to extend to other worlds our own sinfulness."[6] Unlike De Concilio, Pohle "did not project Earth's salvation history onto

---

[4] de Concilio, J. (1889) *Harmony between science and revelation* (pp. 202–232). New York: Fr. Pustet & Co.

[5] Pohle, J. (1889). *Die Sternenwelt und ihre Bewohner* (2nd rev. ed., p. 11). Cologne: Bachem, as quoted in O'Meara, T. F. (2012). *Vast universe: Extraterrestrials and Christian revelation* (p. 85). Collegeville, MN: Liturgical Press.

[6] Ibid, (p. 86)

other planets. 'But even when the evils of sin have infected those worlds it does not follow that an incarnation or redemption must have taken place. God has many other means by which to remit guilt.'"[7]

Like two tennis players attempting to return slicing forehands back across the net, these two-nineteenth-century scholars volleyed their points of agreement—extraterrestrials can and almost certainly do exist; the existence of extraterrestrials poses no existential problems for Catholics or Catholicism—into opposite courts regarding the universality of Christianity and the meaning of Christ's redemptive act on Earth in human form for extraterrestrials. These themes have continued to be both the points of agreement and disagreement throughout the twentieth and into the twenty-first century, though the opinions of most modern Catholic theologians would fit better into Pohle's more broadly inclusive God-can-do-anything domain than into De Concilio's more universally Christian domain.

The twentieth-century scholar against whose work all modern thinking about extraterrestrial life vis-à-vis Roman Catholic theology must be compared is the French philosopher and Jesuit priest Pierre Teilhard de Chardin (1881–1955 C. E.), whose own writings in 1962 were considered full of "serious errors, as to offend Catholic doctrine"[8] by the Sacred Congregation of the Holy Office (the Vatican's doctrinal oversight agency; originally established in 1542 as the Sacred Congregation of the Universal Inquisition, given this name in 1908, merged with the (suppressed) Sacred Congregation of the Index in 1919, and renamed as the Congregation for the Doctrine of the Faith in 1965[9]). Teilhard's writings were suppressed for decades; most of them were unpublished during his lifetime. Despite an admonition from the Holy See as recently as 1981 noting, on the centenary anniversary of the birth of Teilhard de Chardin, that "the judgement given in the Monitum of June 1962" had not been revised,[10] in recent years many of Teilhard's ideas have gained favor at very high levels within Roman Catholicism, including in the writings and speeches of Pope John Paul II and Pope Benedict XVI. In 2009, papal spokesperson, Jesuit Federico Lombardi said, "By now, no one would dream of saying that [Teilhard] is a heterodox author who shouldn't be studied."[11]

Early on in his career Teilhard, who trained as a paleontologist and in the 1920s participated in some of the earliest studies of the approximately 700,000 year-old fossil remains known as Peking Man, fully embraced the idea of human evolution. He articulated a biological/theological argument in which throughout the universe stars and planets

---

[7] O'Meara, T. F. (2012). *Vast universe: Extraterrestrials and Christian revelation* (p. 86). Collegeville, MN: Liturgical Press.

[8] Sacred Congregation of the Holy Office. (1962, June 30). *Warning regarding the writings of Father Teilhard de Chardin*. Retrieved from www.ewtn.com/library/CURIA/CDFTEILH.htm

[9] The duties of the Congregation for the Doctrine of the Faith are "to promote and safeguard the doctrine on the faith and morals throughout the Catholic world: for this reason everything which in any way touches such matter falls within its competence." Oct 25, 2013. Retrieved from http://www.vatican.va/roman_curia/congregations/cfaith/documents/rc_con_cfaith_pro_14071997_en.html

[10] Communiqué of the Press Office of the Holy See. (1981, July 20). *L'Osservatore Romano*. Retrieved from www.ewtn.com/library/CURIA/CDFTEILH.htm

[11] Allen Jr., J. L. (2009, July 28). Pope cites Teilhardian vision of the cosmos as a 'living host.' *National Catholic Reporter.*

form. Subsequently, living things naturally emerge, evolution occurs and intelligence develops among some of those living things. Teilhard concludes that evolution has a direction, "towards the production of increasingly complex systems."[12] He further asserts, in his lecture *Life and the Planets*, delivered in China in 1945, that planets have a significance and importance in the universe far beyond those of stars or gas clouds or galaxies, because in most locations "the Universe is indifferent and even hostile to every kind of life," but it is "on the very humble planets, on them alone, that the mysterious ascent of the world into the sphere of high complexity has a chance to take place. However inconsiderable they may be in the history of sidereal bodies, however accidental their coming into existence, the planets are nothing less than the key-points of the Universe. It is through them that the axis of Life now passes."[13] Ultimately, in Teilhard's cosmological theology, expressed in the pages of his book *The Phenomenon of Man*, life will naturally emerge in many locales in the universe. Then the sphere of intelligence that will be the end-point for evolution on the Earth will "force the bars of its earthly prison, either by finding the means to invade other inhabited planets or ... by getting into psychical touch with other focal points of consciousness across the abysses of space. ... Consciousness would thus finally construct itself by a synthesis of planetary units."[14]

Given a universe full of emergent life forms on multiple planets and one in which evolution plays a vital role, Teilhard forces himself to try to understand evil and Christianity in the universe. In Christian scripture, evil appears for the first time in the form of death, or the threat of death, when God said to Adam, in the Garden of Eden, "but of the tree of the knowledge of good and evil you shall not eat, for in the day that you eat of it you shall die"(Genesis 2:17). In addressing the issue of evil in the universe, Teilhard stakes out a position in which evil does not appear for the first time in the Garden of Eden. In Teilhard's cosmotheology, evil is universal: "If there is an original sin in the world, it can only be and have been everywhere in it and always, from the earliest of the nebulae to be formed as far as the most distant."[15] In his 1920 essay *Fall, Redemption, and Geocentrism*, Teilhard writes, "Either he must redraw the historical representation of original sin; ... or he must restrict the theological Fall and Redemption to a small portion of the universe that has reached such boundless dimensions. The Bible, St Paul, Christ, the Virgin and so on, would hold good only for earth." The basis for his cosmotheology is straightforward: "The *spirit* of the Bible and the Church is perfectly clear: the *whole* world has been corrupted by the Fall and the *whole* of everything has been redeemed." But the fall, which for Teilhard is the appearance of death in the universe, did not occur on the Earth: "... we can no longer derive the whole of evil from one single hominian," he writes. "I must emphasize again that long before man death existed on earth. And in the depths of the heavens, far from any

---

[12] Steinhart, E. (2008). Teilhard de Chardin and Transhumanism. *Journal of Evolution and Technology, 20*, 1–22.

[13] de Chardin, P. T. (1964). *The future of man* (N. Denny, Trans., pp. 103–109). New York: Harper & Row.

[14] de Chardin, P. T. (1959). *The phenomenon of man* (B. Wall, Trans., p. 286). New York: Harper & Brothers.

[15] de Chardin, P. T. (1969). Reflections on original sin. In *Christianity and evolution* (R. Hague, Trans., p. 190). New York: Harcourt Brace Jovanovich.

moral influence of the earth, death also exists. ... in the universe we know today, neither one man nor the whole of mankind can be responsible for contaminating the whole." Original sin, he argues, appears throughout the entire depth and breadth of the universe simply as "the essential reaction of the finite to the creative act."

Given that Teilhard identifies sin as a universal attribute of all that God created, he makes the case that redemption must also be universally available to all of God's creations and that humankind could not have been chosen as the single center for redemption. "The idea of an earth chosen arbitrarily from countless others as the focus of Redemption," Teilhard writes, "is one that I cannot accept." Neither can he accept the idea that Christ's incarnation and redemptive act on Earth is a concept that is of any value to other created beings: "on the other hand the hypothesis of a special revelation, in some millions of centuries to come, teaching the inhabitants of Andromeda that the Word was incarnate on earth, is just ridiculous."[16]

Yet while Teilhard has great faith in the correctness of his cosmotheology, he has no idea how God will work out the details. "All that I can entertain," he concludes, "is the possibility of a multi-aspect Redemption which would be realized, as one and the same Redemption, on all the stars. ... Yet all the worlds do not coincide in time! There were worlds before our own, and there will be worlds after it. ... Unless we introduce a relativity into time we should have to admit, surely, that Christ has still to be incarnate in some as yet unformed star? ... There are times when one almost despairs of being able to disentangle Catholic dogmas from the geocentrism in the framework of which they were born." And so, in the end, Teilhard finds himself with one foot in Thomas Paine's trap, and while he may not be trapped as tightly as De Concilio, he is not completely free of the trap, as is Pohle. While he has escaped the dogmatic trap of original sin, the Christological framework of redemption is one that binds him. For Teilhard, the Earth is not unique and the appearance of Christ on Earth offers no redemptive value for any other beings anywhere else in the universe, past, present or future. Therefore, either "Christ has still to be incarnate in some yet unformed star"—one aspect of Paine's trap—or his Roman Catholic faith applies only to humans. He is not quite willing to accept either of these conclusions.

Teilhard's cosmotheological vision, however incomplete, became acceptable, which is not to say accepted, within Roman Catholic circles after Pope John Paul II addressed the Pontifical Academy of Sciences in 1996. In this presentation, the pope made clear that faithful Catholics may accept evolution, albeit with certain theological qualifications, most importantly that "if the human body takes its origin from pre-existent living matter, the spiritual soul is immediately created by God." In confirmation of his predecessor Pope Pius XII's assertion made in 1950 that there was "no opposition between evolution and the doctrine of the faith," Pope John Paul II said that "new knowledge has led to the recognition of the theory of evolution as more than a hypothesis."[17] If a qualified version of evolution on Earth is acceptable to Pope John Paul II, and therefore to most Roman Catholics

---

[16] de Chardin, P. T. (1969). Fall, redemption, and geocentrism. In *Christianity and evolution* (R. Hague, Trans., pp. 36–44). New York: Harcourt Brace Jovanovich.

[17] Pope John Paul II (1996, October 22). Truth cannot contradict truth. *Address of Pope John Paul II to the Pontifical Academy of Sciences*. Retrieved from www.newadvent.org/library/docs_jp202tc.htm

worldwide, then the natural emergence of life on other worlds is also possible. When working from within such a theological framework, Teilhard's idea for the universal nature of sin becomes acceptable as a solution to one of the fundamental theological issues for the faithful with regard to extraterrestrial life.

Keeping in mind that Teilhard de Chardin was thinking and speaking and writing in the first half of the twentieth century but that his words were largely suppressed by Roman Catholic authorities throughout his lifetime, very little public, Roman Catholic intellectual debate occurred regarding the compatibility, or lack thereof, of extraterrestrial life and the Roman Catholic faith for most of the first half of the twentieth century. In the early 1960s that situation changed, though not dramatically, with the dawn of the space age.

Francis J. Connell (1888–1967 C. E.), an influential American priest who also served as the Dean of the School of Sacred Theology at Catholic University of America from 1949 until 1957, wrote an unusual essay entitled "Flying Saucers and Theology" in 1967. Therein, he writes "it is good for Catholics to know that the principles of their faith are entirely compatible with the most startling possibilities concerning life on other planets."[18]

Father Theodore J. Zubek, in his article "Theological Questions on Space Creatures," published in *The American Ecclesiastical Review* in 1961, echoes the idea that things not mentioned in scripture—extraterrestrial beings—are not precluded by scripture. He writes, "Neither does divine revelation, given to mankind by God, answer the question about the life in space. … The silence of the Bible on the structure of the universe does not exclude the possibility of life outside the earth. … God could have created living beings, including rational creatures, on other planets. … The existence of such beings is not opposed to any truth of the natural or supernatural order."

Zubek then wonders about the souls of such creatures—"In what relation to God would such creatures be?"—and whether they would suffer the sins of Adam. The answer is clear to Zubek, for whom the sins of Adam and the direct inheritance of those sins by all humans is real and happened on Earth: "In any hypothesis, space creatures, not being offspring of Adam, would not belong to the human race and would not have Adam's original sin."[19] As for whether they might be burdened by sin at all, Zubek does not share Teilhard's view of the universal nature of sin. Instead, Zubek firmly asserts that we cannot know the answer, suggesting that extraterrestrials might be free of sin or they might be sinful and that we have no way of knowing their situation in this regard. He writes that these other-worldly beings might "have been put by God into one of several possible states." One of those natural states would be to be free of sin, another would be the fallen state, like man. If fallen, the "space creatures could have been punished by God individually and forever, like fallen angels," or "God could have applied His infinite mercy by simply forgiving the sins of such creatures." Finally, God could demand of these fallen creatures repentance and offer them help, through "a mediator, a redeemer." That mediator, according to Zubek, "would not necessarily have to be one of the Persons of the Holy Trinity." But if God wanted to redeem these creatures, "the possibility of the incarnation

---

[18]Connell, F. J. (1967). Flying saucers and theology. In A. Michel (Ed.), *The truth about flying saucers* (p. 258). New York: Pyramid Books.

[19]Zubek, T. J. (1961). Theological questions on space creatures. *The American Ecclesiastical Review, 145*(6), 393.

of the two other divine Persons, for the purpose of redemption, is not excluded by theologians." Zubek's view is clearly that God can do what God chooses to do, and one of those possible choices is incarnation and a Jesus Christ-like religion, but that is not the only possibility. Roman Catholicism, in Zubek's cosmotheology, is for humans on earth, only, though extraterrestrials could have a similar religion.

Yves Congar, who was appointed as a Cardinal in 1994 and who is recognized as one of the most influential theologians of the twentieth century, attached an Appendix entitled "Has God Peopled the Stars?" to his 1961 monograph *The Wide World is My Parish*.[20] In this essay, he sounds like Galileo—"the intention of the Holy Writ is to teach us how one goes to heaven and not how heaven goes"[21]—when he writes, "Christian doctrine is not concerned with stars or inhabited worlds but with Heaven." He continues, in full agreement with Zubek, saying "Revelation being silent on the matter, Christian doctrine leaves us quite free to think that there are, or are not, other inhabited worlds … Biblically speaking, it is an entirely open question." Just in case his readers missed this point, Congar hammers it home again: "There is no point in being subtle about it. There is nothing in the Bible that touches the matter." He is, however, happy to acknowledge that living beings could inhabit other worlds. "Theology does not see any difficulty about that," he writes.

What is certain, according to Congar, is that "if living beings endowed with understanding exist elsewhere, they too are in God's image, for he is the creator" of everything. For Congar, the natural consequence of being created in God's image is that "They would, then, have a natural religion, analogous to ours, though in forms that we cannot imagine." As for whether these beings "would have been called by God to the life of grace and supernatural revelation, no other answer can be given," he says, "than this: 'It is possible.'" Congar asks whether "there could have been an Incarnation in one of these inhabited worlds? Could the Word of God have taken flesh there and become a Martian, for instance?" In answer to his own questions, Congar thunders that such questions "encroach on a sphere which God reserves to himself." We can never know. We can, however, think about the possibilities, and one such possibility is that other incarnations could happen: "it is not contradictory, and therefore it would not be impossible, that the Word of God, or one of the Persons of the Blessed Trinity, should unite himself to any creature." After all, the God in whom Congar has faith is infinitely powerful. In the end, however, Congar comes back to Earth and reiterates that the only thing that matters for us is "God's actual plan where we are. And for us this plan has only one name: Jesus Christ." Ultimately, Congar's theology, like that of Zubek, is very clear. Roman Catholicism is for humans on Earth, not for living beings on Mars or Venus or anywhere else, though Catholic-like realities (including sin, incarnation and resurrection) are possible elsewhere.

The German Jesuit Karl Rahner (1904–1984 C. E.), another leading Catholic theologian of the twentieth century, was also one of the few who had something to say on the subject of extraterrestrial life: "This question must be raised," he writes in words that mirror those of Teilhard de Chardin, "in view of the vast immensity of the material cosmos as a world coming to be. If we imagine the cosmos as a world coming to be … then it is really

---

[20] Congar, Y. (1961). *The wide world is my parish*. Baltimore: Helicon Press.

[21] Galilei, G. (1989). Letter to the Grand Duchess Christina, 1615. In M. A. Finocchiaro (Ed.), *The Galileo affair* (p. 96). Berkeley: University of California.

not to be taken for granted that this aim has been successful only at the tiny point [in the cosmos] we know as our earth."[22] Given that those extraterrestrials almost certainly exist, Rahner asserts that God, his Roman Catholic God, will be involved in their lives. He writes that "the material cosmos as a whole, whose meaning and goal is the fulfillment of freedom, will one day be subsumed into the fullness of God's self-communication to the material and spiritual cosmos, and that this will happen through many histories of freedom which do not only take place on our earth."[23]

Allowing that extraterrestrial beings could exist and that a universal God could communicate with those beings is one thing. Understanding the relationship of those beings to the God in whom Roman Catholics place their faith is another thing entirely. Do these other beings require salvation? If so, what form does salvation take for them? Voicing similar views to those of Congar and Zubek, Rahner argued without any equivocation that the incarnation of Christ on Earth was of and for the salvation of humans and only humans: "A theologian can hardly say more about this issue than to indicate that Christian revelation has as its goal the salvation of the human race; it does not give answers to questions which do not in an important way actually touch the realization of this salvation in freedom."[24]

Rahner accepts that the God of Roman Catholicism is the God of the entire universe but that the Roman Catholic religion, which incorporates the ideas of the incarnation and resurrection of Jesus at a particular time and place on Earth, is meant only for humans on Earth. In Rahner's cosmotheology, other intelligent beings, if they exist, will receive God's love; but, he opines, we should make no assumptions as to how good and evil play out on their worlds or about how God offers his love to other creatures. Despite his extreme openness to how religions on other worlds might be constructed, he continues: "It cannot be proved that a multiple incarnation in different histories of salvation is absolutely unthinkable."[25] Again sounding like Congar and Zubek, he suggests that multiple incarnations are not necessary but are possible.

Thus, Rahner and Congar and Zubek have each set one foot firmly on both sides of the issue. On the one hand, they are extremely open to the possibility that Christian revelation is only about the salvation of humans and not at all about the salvation of extraterrestrials. Nevertheless, they are all eager to suggest that the same path to salvation (multiple incarnations, the Word made flesh) could and might occur anywhere and everywhere: the active, all-powerful God in whom they believe could choose to use the same method to offer salvation to extraterrestrial beings as God chose to offer to humans. The theological rationalizing that 'allows,' though does not demand, the possibility that the same Christological framework can occur over and over again in the universe is ultimately a mapping of Christianity onto the larger universe. In the end, the very idea that multiple incarnations

---

[22] As quoted in Fisher, C. L., & Fergusson, D. (2006). Karl Rahner and the extra-terrestrial intelligence question. *Heythrop Journal, 47,* 275.

[23] Rahner, K. (1978). *Foundations of Christian faith* (pp. 445–446). New York: Seabury Press.

[24] As quoted in O'Meara, T. F. (1999). Christian theology and extraterrestrial intelligent life. *Theological Studies, 60,* 19.

[25] As quoted in Fisher, C. L., & Fergusson, D. (2006). Karl Rahner and the extra-terrestrial intelligence question. *Heythrop Journal, 47,* 275.

might occur begins to sound like wishful thinking that the basic theological framework of Roman Catholicism might be universal, and this thinking places Rahner, Congar and Zubek in solidarity with Teilhard de Chardin and also at least partly in the camp of those ridiculed by Thomas Paine.

Writing in the journal *Theological Studies* in 1999, Thomas O'Meara advises, in a vein similar to that of Rahner, Congar and Zubek, that we have no reason to assume that our ideas of sin and salvation and our experience with salvation via the personhood of Jesus, apply to extraterrestrials. "At no point in the Hebrew or Christian Scriptures do we learn that there is another race elsewhere in the universe, or that there is not."[26] He continues, making the point that "It is superficial and arrogant to assert that the Christian or Jewish revelation of a wisdom plan for salvation history on earth is about other creatures. Faith affirms that the Logos (the concept of Christ, the Second Person of the Trinity, as the Word of God) has been incarnate on a planet located, in past Ptolemaic astronomy, in a small, closed system. The Logos, the second person of the divine Trinity, indeed has a universal domination, but Jesus, Messiah and Savior, has a relationship to terrestrials existing within one history of sin and grace."

Again, sounding like Congar, O'Meara suggests that we gain nothing, certainly we learn nothing about ourselves or our relationship with God, by worrying about the existence or non-existence of evil on other inhabited worlds or concerning ourselves with the relationship God will or will not establish with those beings. "No matter how thankful human beings are for special contacts with God," he writes, "they do not honor revelation by projecting terrestrial religion (and its context of proneness to evil and sensuality) beyond earth and thereby limiting divine wisdom and power. ... The history of sin and salvation recorded in the two testaments of the Bible is not a history of the universe; it is a particular religious history on one planet... the central importance of Jesus for us does not necessarily imply anything about other races on other planets."

O'Meara does, however, conclude that the Holy Trinity of the Father, Son and Holy Spirit is universal, that salvation via some kind of incarnation might be necessary for other creatures, and thus something akin to Roman Catholicism could be universal: "The divine generosity that led once to the Incarnation on earth suggests that there might be other incarnations—many incarnations and in various species, many creatures touched in one or another special, metaphysical way by a person of the Trinity." In this way, he is a bit more assertive about the universality of certain aspects of Christianity, than are Rahner, Congar and Zubek. O'Meara's cosmotheological approach, one in which anything is possible and anything is allowed in the universe but in which a Roman Catholic-like Trinitarian religion, one that involves sin and incarnations and salvation, is possible—and by subtle inference is desirable—appears to represent the direction in which Roman Catholicism thought has moved since the middle of the twentieth century.

John F. Haught is one of the few Roman Catholic theologians who in recent years have written extensively about the theological questions associated with the possible discovery of extraterrestrial life. His essay "Theology after Contact: Religion and Extraterrestrial Intelligent Life" was published in 2001 in the *Annals New York*

---

[26] O'Meara, T. F. (1999). Christian theology and extraterrestrial intelligent life. *Theological Studies, 60*, 20.

Academy of Sciences.[27] In this essay Haught, a Senior Fellow at the Woodstock Theological Center and the former Chair of the Department of Theology at Georgetown University, wrote, "Theology's relevance to SETI lies most fundamentally in its conviction that all possible worlds have a common origin in the one God." He then provides a modern answer to a question posed nearly 750 years ago by Thomas Aquinas in his *Summa Theologiae* as to why God created such a diversity of living things: the diversity of life is an expression of the greatness of God. "The basic theistic belief that the reality of God has already become partially manifested in the extravagant multiplicity of non-living and living beings on our own planet should already have prepared the religious mind for a disclosure of even richer diversity elsewhere." With this in mind, the existence of extraterrestrial life forms certainly offers nothing that could threaten Roman Catholic religiosity. In fact, in Haught's view, the existence of extraterrestrial Others offers the opposite of a threat: evidence of God's greatness.

Haught holds the opinion that humans are special—"we should emphasize that it is biologically inconceivable that there would be other *humans* anywhere else in the universe; so our uniqueness as a species is virtually guaranteed." He also holds the view that humans are neither superior nor inferior in comparison to any other sentient beings. "We express our own unique human dignity and value not by looking for signs of our mental and ethical superiority over other forms of life, but by following a path of service and even self-sacrifice with respect to the whole of life, wherever it may be present ... Thus, it is inconceivable that the eventual encounter with beings that may in some ways be our superiors would ever render such instruction obsolete." As for whether humans have been "specially called or set apart by God," Haught's answer is no. Any sense of 'special election' must be understood to mean only that humans have "a vocation to serve the cause of life and justice" and that vocation does not lift us "out of our fundamental relatedness to the entire cosmic community of beings."

Haught articulates his belief that extraterrestrials are likely to be religious in nature because they are biological beings with finite life spans. "They too would be subject to transience and eventual perishing. They, like us, would be subject to the threat of failure, and eventually nonbeing ... these Others ... would also be in search of ways to transcend the limits on their particular forms of life." As for the universality of Christianity, he argues that what is fundamental about Christianity has little to do with the birth, death or resurrection of Jesus. "What seems to be universally applicable in Christianity (and indeed other religious traditions)," he writes, "is the ideal of embracing rather than eliminating diversity." In doing so, he expects that Christianity will be enriched rather than dissolved and that in the future, "solidarity with Christ would continue for the Christian to mean belonging to one whose own life was itself a vulnerable openness to the estranged and alien, to what does not yet belong. After contact, 'belonging to Christ' could then readily be thought of as requiring a more radical inclusiveness than before, one open to and supportive of the adventures of many intelligent worlds." Furthermore, Haught believes that the discovery of intelligent extraterrestrials would lead us to recognize that life is a gift

---

[27] Haught, J. F. (2001, December). Theology after contact: Religion and extraterrestrial intelligent life. *Annals of the New York Academy of Sciences, 950,* 296–308.

from the creator, not an accident. "Certainly the existence of ETI would force us to reexamine the claim by evolutionists … that life and intelligence are the results of utterly improbable, purely random statistical aberrations in an overwhelmingly lifeless and mindless universe. In this respect, SETI would seem to have theological importance." This belief in directed evolution and the idea of cosmic purpose echoes that expressed earlier by Teilhard de Chardin. Thus, for Haught, the discovery of and, even better, contact with extraterrestrial beings will strengthen our understanding of and belief in God and, for him, in Christianity; but, the theological issues of original sin and redemption are simply nonissues for him. In his view, they don't even merit discussion as part of our collective efforts in trying to understand extraterrestrial life. We do not need to worry, he implies, about how religion will be manifest for those Others. From this, we can conclude that Haught's view is that Roman Catholicism is a terrestrial religion, God's gift to humanity. His views are also not mainstream Roman Catholicism.

Modern Roman Catholic astronomers, in following in the footsteps of their most famous scientific forebear, have also entered into this conversation. After all, Galileo, who started us down this road toward public debates about the relationship between science and religion, took the position that "God reveals Himself to us no less excellently in the effects of nature than in the sacred words of scripture."[28]

Guy Consolmagno is a professional astronomer who earned his Ph.D. in Astronomy from the University of Arizona, served as Chair of the Division of Planetary Sciences of the American Astronomical Society (2006–2007), and was President of the International Astronomical Union's Commission 16 on Moons and Planets from 2003 until 2006. His credentials as an astronomer are in order. He entered the Jesuit order in 1989, took his vows in 1991 and accepted an assignment with the Vatican Observatory in 1993. At this time, he is curator of the Vatican's extensive meteorite collection at Castel Gandalfo.

In his book *Brother Astronomer: Adventures of a Vatican Scientist* (2000), he writes, "It is not just humankind, but the whole of creation, that was transformed and elevated by the existence of Christ."[29] Exactly what Consolmagno means by Christ is unclear. Does he mean the Risen Jesus, the man known as Jesus of Nazareth who according to Christian belief was resurrected and arose from the dead, or does he mean the Word of God present in the Trinity who is also present in Jesus of Nazareth? Whatever he means, he continues, "Finding any sort of life off planet Earth, either bacteria or extraterrestrials, would pose no problem for religion." Roman Catholicism, Consolmagno asserts, would embrace the discovery of extraterrestrial life, but somehow, in someway, every form of life everywhere in the universe will be "elevated" by Christ. In at least this sense, Consolmagno's view is that important aspects of Christianity are universal. He continues, noting that "God created the whole universe. There's nothing that makes one place more special than another. Religious people have been able to think in these 'cosmic' terms all along, and happily speculated about 'other worlds' long before the science fiction crowd had adopted the concept."

[28] Galilei, G. (1989). Letter to the Grand Duchess Christina, 1615. In M. A. Finocchiaro (Ed.). *The Galileo affair* (p. 93). Berkeley: University of California.

[29] Consolmagno, G. (2000). *Brother astronomer: Adventures of a Vatican scientist* (pp. 106, 150–152). New York: McGraw-Hill.

Consolmagno continues, effectively ignoring Haught's advice and worrying about sin and incarnation, writing "But there is one crucial question that will face Christianity if, or when, extraterrestrial intelligence is discovered. That's the question about what the Incarnation means to other species. In other words, would aliens need to have their own version of Jesus? Do aliens need to be saved? Depends if they are subject to 'original sin' or not. The traditional theology of original sin, tracing it back to the origins of the human race, says absolutely nothing about other entities, either way. Once we find other intelligences, we'll be in a better position to expand that theology."

Consolmagno's position here is that there is no sense dealing with the hypothetical question about original sin, even though he clearly thinks that the question of original sin for extraterrestrials is one that is worth thinking about. That problem can and will be solved if and when we encounter it. For now, for him, the issue is important but at this time no answer to that question can be meaningful. After dismissing the reasonableness of pursuing this topic any further, he then is willing to deal with half of the hypothetical: what if the extraterrestrials do suffer from original sin? "Assuming that original sin, the problem of evil, does face other intelligences," he writes, "what role does Christian salvation play in their world?" Consolmagno provides a firm answer, that Christian salvation applies to the entire universe: "St. Paul's hymns in Colossians 1 and Ephesians 1 make it clear that the resurrection of Christ applies to all creation ('everything in the heavens and everything on earth'). It is the definitive salvation event for the cosmos. Another bit of Biblical evidence is the opening of John's Gospel, who tells us that The Word (which is to say, the Incarnation of God) was present from the Beginning; it is part and parcel of the woof and weave of the universe." Thus Consolmagno, after asserting that we can know nothing about sin and salvation for extraterrestrials, claims to know everything that needs to be known. His answer is that the redemption event—manifest as Jesus on the cross—occurred only once in the history of the universe, that that event occurred here on Earth, and that the resurrection of Christ applies to all creation and is the definitive salvation event for the cosmos.

As for how knowledge of the resurrection could be conveyed across the vastnesses of space and time, Consolmagno's simplistic answer is that he doesn't know how, but knows that it will be done: "Just how this 'Word' might be 'spoken' to the rest of the intelligent universe, I don't know. But it will be in 'words' (that is, events) appropriate to those beings. … The point here is that, even though the life of Jesus occurred at a specific space-time point, on a particular world line (to put it in general relativity terms), it also was an event that John's Gospel describes as occurring in the beginning—the one point that is simultaneous in all world lines, and so present in all time and in all space. Thus, there can only be one Incarnation—though various ET civilizations may or may not have experienced that Incarnation in the same way that Earth did."

Consolmagno recognizes that "We have no data about other, nonhuman civilizations. They may not even exist; or they may be plentiful." And if they do exist, "ETs may not be aware of the idea of an Incarnation, or they may have their own experience of the matter. Their experience may be so alien from ours that even though they have experienced God in their own way, it's an experience that we will never be able to share, nor they share in our experience."

What should we make of Consolmagno's words? In his view, the redemption of Christ provides salvation for all of creation and thus some version of Christianity is fully universal, even if the aliens do not know about and understand Christ in the same manner with which Roman Catholics know and understand him. In addition, Consolmagno assumes that the concepts of sin and original sin exists everywhere, though he does not assume that all aliens are sinful. Any particular species of aliens that is sinful must suffer its own version of original sin. Those aliens would require an incarnation for redemption, and the incarnation of Christ on Earth would be for them, also, though they need not even know about Christ in order to benefit from his redemptive act on Earth. On the other hand, any particular species of aliens that is not sinful did not suffer original sin and would not require an incarnation for redemption. Nevertheless, Christ still exists for them.

This view places humankind at a singular, central place in the religious life of the universe. The incarnation of Christ occurred here, on our planet. This all-important event in the history of the universe, the event that offers salvation to any and all sinful species anywhere and everywhere in the universe, took place right here in our backyard, only 2,000 years ago. With this all-encompassing, universal Christianity, Consolmagno returns us at least partially to pre-Copernican times when the Earth and humankind held the central place in God's creation.

José Gabriel Funes is a Jesuit priest and a professional astronomer. He holds a Ph.D. in Astronomy from the University of Padua and has served as the director of the Vatican Observatory since 2006. In an interview in 2008, published in the *L'Osservatore Romano*,[30] he asserts that extraterrestrial life could exist—"As a multiplicity of creatures exist on earth, so there could be other beings, also intelligent, created by God." He also expresses the view that "This does not contrast with our faith because we cannot put limits on the creative freedom of God. To say it with Saint Francis, if we consider earthly creatures as 'brother' and 'sister,' why cannot we also speak of an 'extraterrestrial brother?'"

As soon as Funes acknowledges that 'extraterrestrial brothers' could exist, he takes on the existential questions at the heart of Roman Catholicism: Are they sinful? If so, how will they be saved? Funes suggests that if they do exist, they might be unlike humans, without sin. Humans are the lost sheep, the ones who have strayed from the herd: "We borrow the gospel image of the lost sheep. The pastor leaves the 99 in the herd to go look for the one that is lost. We think that in this universe there can be 100 sheep, corresponding to diverse forms of creatures. We that belong to the human race could be precisely the lost sheep, sinners who have need of a pastor. God was made man in Jesus to save us. In this way, if other intelligent beings existed, it is not said that they would have need of redemption. They could remain in full friendship with their Creator."

Funes's initial answer, concerning extraterrestrials already full of grace, is the easy one. When pushed by his interviewer as to whether, if the extraterrestrials were sinners, they could enjoy redemption, Funes answered, "Jesus has been incarnated once, for everyone. The incarnation is an unique and unrepeatable event. I am therefore sure that they, in some way, would have the possibility to enjoy God's mercy, as it has been for us men." Thus, in Funes's view God is universal, we humans are special and Christianity is specifically for and about us. The incarnation happened once, here, on Earth, where we sinful

---

[30] The extraterrestrial is my brother. (2008, May 14). *L'Osservatore Romano*.

humans live. God will extend his mercy to those other creatures, but will not send His son to bear witness on the cross on any other planets or in any other places in the universe. If those other beings are sinful, God will find a means to redeem them that does not involve knowledge about His Son dying on a cross on Earth and might not even involve incarnation.

While the Roman Catholic Church is monotheistic in faith, and while it has a strong administrative hierarchy with a pope who can speak for all of the faithful, it is not without differences of opinion on the subject of how extraterrestrial life should be understood with respect to Catholicism. One of the more conservative of the modern viewpoints is that of theologian and bishop Joseph Augustine Di Noia.

In an interview with the *National Catholic Reporter* in 2004,[31] Di Noia, then under-secretary of the Congregation for the Doctrine of the Faith and since 2012 also vice-president of the Pontifical Commission "Eccleisa Dei," started off by offering the now-traditional approach toward understanding extraterrestrial life, "Christians have always understood that the entire cosmos is a creation of God, that any life anywhere is a divine creation." That is, if scientists discover extraterrestrial life, Roman Catholicism will embrace that discovery in full. Di Noia followed up this warm embrace by asserting a fully Christ-centered theology, "If there are other persons in the universe, we can at least say that they too are involved in the same divine plan and are intended to share in the Trinitarian communion of life." For Di Noia, something similar to Roman Catholicism is the singular, universal religion, though Di Noia makes no effort to solve the Thomas Paine problem. Di Noia's viewpoint is fairly conservative, but his is not a singular voice from within the Roman Catholic community.

Australian Gerald O'Collins, a theologian who taught at the Jesuits' Gregorian University in Rome from 1973 until 2006, remarked, in the same *National Catholic Reporter* article in 2004, "I don't think the discovery of life on other planets would pose a qualitatively different challenge than the discovery of the New World. … we survived that, and in the end it deepened our understanding of Christ as a truly universal savior."[32] For O'Collins as for Di Noia, Christ is a "universal savior." Such a Christology avoids asking and answering the theological questions as to why these other beings need a savior as well as how knowledge of Christ is transmitted across the incomprehensible depths of space and time. In addition, if the discovery of life on other worlds will be qualitatively similar to the discovery by Europeans of non-Christian humans in the New World, we should ask what the qualities of that discovery were. Five hundred years ago, for the Roman Catholic Church the challenge of the discovery of the New World was the challenge of teaching the newly-discovered human beings about Jesus Christ and converting them to Catholicism. Though O'Collins' meaning is not absolutely clear, one can understand his words to mean that in the future, should we discover extraterrestrial beings, transmitting knowledge about Jesus Christ and about the Roman Catholic faith to beings throughout the universe will become the job of the Roman Catholic Church. He would be offering instructions to those Jesuit priests headed for Rakhat. If so, his ideas about proselytizing to the entire universe are quite far outside of the mainstream of modern Catholicism.

---

[31] Allen Jr., J. L. (2004, February 27). This time, Catholic Church is ready. *National Catholic Reporter*.
[32] Ibid.

Both Di Noia and O'Collins suggest that "some theological tinkering" may be necessary, "especially with the concept of original sin. How can persons on other planets share in the stain of guilt derived from Adam and Eve, from whom they are most probably not descended? Yet if they don't [share in the guilt of original sin], what exactly is the condition from which Christ has redeemed them?" O'Collins concluded the interview, saying, "We'd have to work on it a little bit, I suppose. But anyone who thinks the doctrine of original sin is more important than Jesus being the universal savior already has their priorities out of line."

For Di Noia and O'Collins, the concept of original sin might not be universal, but the concept that all life forms in the universe require redemption by a savior is universal. Furthermore, in their view only one such savior, Jesus Christ, exists across the vast length and breadth and temporal expanse of the universe, and only Roman Catholicism provides the spiritual path to salvation.

Roch Kereszty (b. 1933), a priest at the Cistercian Abbey of Our Lady of Dallas and a theologian at the University of Dallas, added an appendix entitled "Christ and Possible Other Universes and Extraterrestrial Beings" to his book *Jesus Christ: Fundamentals of Theology*.[33] Kereszty emphasizes that "the existence of other universes or extraterrestrial intelligent beings is theologically possible but not more probable than their non-existence." This view is consistent with those of other modern Roman Catholic theologians. Kereszty then nearly dances off a theological cliff when he asserts that God creates "out of love for the creature" and that, therefore, if God chose to create any extraterrestrials, "his purpose is that the intelligent creature should share in some way or other in his own joy." Any alien intelligences will have been created "in the image of the Son of God ... because they are endowed with intellect and freedom," and any such beings would be "in some sort of communion with God the Father through the Son." Though he concludes that God has "absolute freedom," Kereszty seems to have placed significant constraints on what God can and cannot choose to do. In Kereszty's cosmotheology, he seems to understand exactly what God did or has done and precisely why God does what God does; God's possible actions throughout the universe appear identical to how Kereszty understands God's actions as they relate to humanity, and they look like a one-to-one mapping of Roman Catholicism onto the rest of the universe. Perhaps, as in the way Thomas Aquinas explained that Aristotle's laws of physics were a description of the universe God chose to create and not a constraint on the universe God was permitted to create, Kereszty is only describing the religious options God chose to place in the universe for all intelligent beings and is not limiting God's authority for creating intelligent beings who might not be made in the image of the Son of God or who might not be in communion with God the Father through the Son. But it doesn't sound that way.

In 2007, Ilia Delio, of the Order of Saint Francis and a scholar of Franciscan theology at Georgetown University, offered another conservative point of view. Delio writes that Catholic "theology is integrally related to cosmology. Once cosmology changes, so too does theology." Delio notes that "Speculation on the meaning of Christ for extraterrestrial (ET) life has received little attention in the science and religion dialogue, despite advances

---

[33] Kereszty, R. A. (1991). Christ and possible other universes and extraterrestrial beings. In *Jesus Christ: Fundamentals of theology* (pp. 376–381). New York: Alba House.

in astronomy, astrobiology and space exploration. Perhaps the hesitation in undertaking this pursuit is the fear of disrupting the core doctrine of Christian faith, namely, the work of Jesus Christ."[34]

Could the discovery of extraterrestrial life disrupt the core doctrine of Christianity, as Delio asserts might happen? Not as she understands Christianity in the first decade of the twenty-first century. Delio presents a theology strongly based on the ideas of medieval thinkers such as Saint Bonaventure (1221–1274 C. E.) and Duns Scotus (1266–1308 C. E.): "*It is shown* that Incarnation takes place wherever there is intelligible life."[35] Delio asserts "all possible worlds, created through the Word of God, bear a spiritual potency within them and are open to spiritual transformation." The assumptions that underlie Delio's assertion is that all intelligent beings on all other worlds require spiritual transformation, that those transformations will occur through some form of incarnation, and that those incarnations will be unique events for each world. "Because Incarnation may take on other extraterrestrial life forms, it is suggested that there may be multiple incarnations but only one Christ." By concluding that the one Christ will be continually incarnated on other planets, Delio offers a twenty-first century viewpoint that Thomas Paine would be pleased to ridicule. Hidden within this salvation-by-other-means theology are the fundamental ideas that evil is universal (and thus is independent of the Earth and the Garden of Eden), that for God "creation and incarnation belong together," and that "incarnation completes that which God creates." Of course, Christianity is not complete without salvation, so for Christians the incarnation is necessarily followed by the crucifixion and resurrection. Delio, though, never discusses the issue of the resurrection of Christ on other worlds or how salvation for extraterrestrials is achieved through incarnation alone. Delio's position has a bit of Teilhard de Chardin's view that evil exists throughout the universe as a necessary consequence of God's creative activity, though Delio does not offer an explanation for the universal origin of evil. Why God must employ incarnation as a means for salvation (and how the salvation of extraterrestrials happens through incarnation alone, since this is not how salvation of Christians is gained on Earth) everywhere is also a mystery Delio leaves unexplained, but that need, she asserts, is fundamental to life. Extraterrestrials, therefore, "if they exist, will be open to an incarnation and, in the broadest sense, 'saved'" and that "on an ET level, incarnation must assume a form that includes the material reality of that creation, in whatever way that creation is constituted."

With a very few exceptions, the views expressed by Roman Catholic theologians have been and continue to be that the ideas of creation and salvation apply to all sentient beings in the universe and that these ideas come from an active Trinitarian God. Neither the existence nor the absence of extraterrestrial life would challenge this view. From a Roman Catholic perspective, if sentient extraterrestrials exist some but perhaps not all such species may suffer original sin and will require redemption. Scholars disagree concerning the need for and the means of redemption for those Others in need. In the views of some, the singular redemptive act that occurred on Earth 2,000 years ago has already triggered the salvation of all creatures everywhere, and time is not a barrier to the passage of this

---

[34] Delio, I., O. S. F. (2007). Christ and extraterrestrial life. *Theology and Science, 5*, 3.

[35] Ibid, [emphasis added].

salvation knowledge throughout space and time; by implication, the physical Earth and humanity are spiritually central to the salvation story of the entire universe. A more widespread Roman Catholic view is that God has the power to offer redemption to Others with no preconditions, but that extraterrestrials likely will be sinful and in need of redemption. The Others may or may not experience a local incarnation of Christ; they may be given knowledge of the incarnation and redemption of the Son of God on Earth.

A recent statement by Pope Francis (b. 1936 C. E.) may suggest that an old theological worldview could help define a new theological universe-view for Roman Catholics. In 2013, he said, "The Lord has redeemed all of us, all of us, with the Blood of Christ; all of us, not just Catholics. Everyone."[36] Pope Francis clearly intends these words to be about Hindus, Jews, Muslims and Buddhists, and most likely did not have extraterrestrials in mind. According to Pope Francis, none of these non-Christians humans need to convert to Christianity in order to be redeemed. Hindus, Jews, Muslims and Buddhists might not be concerned with whether the Roman Catholic pope thinks that Jesus on the cross redeems non-Catholics. If, however, the pope were to extend this idea to extraterrestrials, then this humancentric and geocentric idea could allow Catholics to believe that any not-yet-discovered extraterrestrials, beings who might never know about Jesus and therefore can neither embrace nor reject Christianity, could nevertheless be saved through mediating powers brought to bear in the life, death and resurrection of Jesus Christ. While Francis's salvation theology leaves unexplained how or why any of this happens or needs to happen, if applied to the entire universe it would be a declaration by the leader of Roman Catholicism that in his view the God of the Roman Catholic faith is the God of the entire universe and that Roman Catholicism itself is the one true religion for the entire universe.

---

[36] Pope Francis, quoted in *Heaven for atheists? Pope sparks debate*. Retrieved May 23, 2013, from http://religion.blogs.cnn.com/2013/05/23/heaven-for-atheists-pope-sparks-debate/

9

Orthodox Christianity

For I will behold Thy heavens, the work of Thy fingers;
the moon and the stars, which Thou has founded.
What is man that Thou art mindful of him?

Psalm 8 (From *Orthodox England on the 'net.*
Retrieved from http://orthodoxengland.org.uk/hp.php)

Through many statements, the Ecumenical Patriarch of Constantinople Bartholomew I (born 1940; installed as Patriarch in 1991), the spiritual leader of Orthodox Christianity, has repeatedly emphasized that the universe is entirely about humanity, the community of sentient beings living on a single planet called Earth. The universe is not about the lady-bugs or the raccoons, let alone any possible Martians or Alpha Centaurians. The universe exists for humanity alone, and the earth is the single place in which human actions play out. As spiritual beings, we are alone. The more extreme viewpoint of certain vocal, fun-damentalist Orthodox Christians is that extraterrestrials almost certainly exist and that these aliens are spiritual demons doing the work of Satan.

During the foundational years for Christianity, during the first several centuries after the time of Jesus, hundreds of independent churches were founded. Five of these churches ultimately emerged and were recognized and empowered by Church councils as the stron-gest independent leaders of the Christian movement. According to tradition, though not clearly supported by the historical record, these five churches were founded or governed by one or more of Jesus's apostles, and so are known as apostolic churches. These churches were in Alexandria ('founded' by Saint Mark), Antioch (moved to Damascus, Syria, in the fourteenth century; 'founded by Saints Paul and Peter'), Constantinople (in modern Istanbul; 'founded' by Saint Andrew), Jerusalem ('founded' by Saints Peter and James) and Rome ('founded' by Saints Peter and Paul). After the Roman empire fractured and collapsed during the fourth and fifth centuries, the Church of Rome was isolated by

D.A. Weintraub, *Religions and Extraterrestrial Life: How Will We Deal With It?*,
Springer Praxis Books, DOI 10.1007/978-3-319-05056-0_9,
© Springer International Publishing Switzerland 2014

distance, politics and language from these other four apostolic churches, as it was the only one of these five churches located in the Latin-speaking, western part of what was the Roman empire. The other four churches found themselves in the Greek-speaking remnants of the eastern empire.

During the fourth through the eleventh centuries, several major differences in doctrine and practice generated tension between the leadership of the Church of Rome and the leaders of the several Greek churches. Among the most important differences were the decision to exclude (Greek) or include (Latin) the words "and from the Son" in the Nicene Creed, the use of leavened (Greek) or unleavened (Latin) bread for the sacrament of the Eucharist and the practice of using (Greek) or not using (Latin) icons as sacramental devices for deepening the spiritual contact between the human and the divine. In 1054 C. E., the pope in Rome, Leo IX, and the patriarch of Constantinople, Michael Cerularius, mutually excommunicated each other. These acts of excommunication, which remained in place for more than 900 years, lifted only in 1965 C. E., triggered the Schism of 1054. This schism establishing a permanent split between the Roman Catholic Church, headquartered in Rome and led by the pope, and the Eastern Orthodox Church, which then had and continues to have many independent churches and no single leader. All the Eastern Orthodox churches, however, respect the Ecumenical Patriarch of Constantinople as the *primus inter pares* (the first among equals), the spiritual leader of the world's Orthodox Christians.

Today, about 300 million people around the world are members of self-governing Orthodox Christian churches, including the Church of Constantinople (headquarters in Istanbul, Turkey), the Church of Antioch (headquarters in Damascus, Syria) and the Churches of Jerusalem, Russia, Serbia, Romania, Bulgaria, Georgia, Cyprus, Greece, Poland, Albania, America, the Czech and Slovak republic, Sinai, Crete, Finland, Japan, China and Ukraine. Eastern Orthodox teachings derive from both Holy Scripture and Sacred Tradition, with the latter being words and beliefs transmitted orally by the apostles. The Bible itself is considered by Orthodox Christians to be the divinely inspired Word of God. The written and oral teachings are valued equally and are understood and interpreted by Ecumenical Councils.

For many years, Bartholomew I has "preached that caring for the environment is a religious imperative."[1] In his many speeches on the topic of the environment, the patriarch has dropped hints about his views concerning life in the universe. He has written that "the entire universe participates in a celebration of life,"[2] which could be understood as accepting the possibility that humankind shares the universe with living beings, whether single-celled or advanced life forms, that live beyond the Earth.

Such an understanding, however, would not be consistent with some of Bartholomew I's other statements, in which he appears to suggest that God created the entire universe exclusively for humankind. For example, in his 1990 essay *Orthodox Perspectives on*

---

[1] Simons, M. (2012, December 3). Orthodox leader deepens progressive stance on environment. *New York Times*.

[2] Bartholomew, I. *Address at the Environmental Symposium, Santa Barbara, California*. Retrieved from http://patriarchate.org/documents/santa-barbara-symposium

Creation, he writes, "The value of the creation is seen … in the fact that it is appointed by God to be the home for living beings … to be the context for God's Incarnation and humankind's deification."[3] And in his 1998 *Message for the Day of the Protection of the Environment*, he says, "Even though the human being … is only a miniscule speck in the face of the immense universe, it is a fact that the entire universe is endowed with meaning by the very presence of humanity within it. Based on this assurance, even leading contemporary scientists accept that the universe is infused with the so-called 'human principle,' meaning that it came about and exists for the sake of humanity."[4] While the evidence offered by the patriarch that leading contemporary scientists have endorsed the idea that the universe exists for our sake could be called into question, he firmly stands behind this claim. He even repeated this general idea in his *Vespers for the Protection of the Environment*, issued in 2001, saying "The quintessence of this overflowing love of God … was the boundlessly abundant, personal, spiritual, and body-spirit creation of angels and people, and the infinite creation of the whole universe to serve them."[5]

One might conclude that from Bartholomew I's perspective on the Orthodox tradition that no reason exists for God to have created extraterrestrial life. Certainly, the Orthodox Christian view, as expressed by the patriarch, is that the universe is purposeful and that extraterrestrial life could only exist if by existing it promotes the sole purpose of the universe, which is for humans to exalt God and seek God's approval. Since it appears to be self-evident that extraterrestrial life would not help humans exalt God, an Orthodox Christian could justifiably conclude that God had no reason to create life beyond the Earth. Therefore, no life should exist beyond the Earth.

In their 2008 essay "An Eastern Orthodox Perspective on Microbial Life on Mars," published in the journal *Theology and Science*, A. Randall Olson, the head of the Department of Environmental Sciences at Nova Scotia Agricultural College, and Vladimir V. M. Tobin, the rector of the St. Vladimir Orthodox Church in Halifax, Nova Scotia, assert that "there is, to date, very little information available concerning an Eastern Orthodox Christian perspective" on extraterrestrial life. "To be sure," they wrote, "there is nothing in Orthodox theological tradition that would affirm the possibility of life on other planets."[6] Olson and Tobin then make the case that extraterrestrial microbial life—they very pointedly avoid addressing the issue of more advanced extraterrestrial life forms—could exist without contradicting Orthodox theology.

From a biological perspective, they argue that the evolutionary imperative is strong; therefore, if microbes exist on another planet, such microbes, given enough time and a sufficiently supportive environment—enough food, enough energy, a chemically rich and

---

[3] Bartholomew, I. (1990). Orthodox perspectives on creation, an excerpt. In G. Limouris (Ed.) *Justice, peace and the integrity of creation: Insights from orthodoxy*. Geneva: WCC Publications. Retrieved from http://patriarchate.org/tbd/theological-perspectives/praxis-orthodox-perspective

[4] Ecumenical Patriarch Bartholomew I "Message for the Day of the Protection of the Environment". (1998, September 1). Retrieved from http://patriarchate.org/documents/1998-encyclical

[5] Bartholomew, I. (2001). *Foreword for the English translation of "Vespers for the Protection of the Environment"*. Retrieved from http://patriarchate.org/documents/protection-environment-2001

[6] Olson, A. R., & Tobin, V. V. M. (2008). An Eastern orthodox perspective on microbial life on Mars. *Theology and Science, 6*(4), 422–437.

active medium in which they live, not too much damaging radiation—will change and some of those changes may lead to more advanced life forms. Can intelligent life not evolve, assuming life emerges at all? Up until the present time, this evolutionary imperative has not mattered for Orthodox Christian theologians because Orthodox Christianity has not yet made peace with modern science, including evolution. In such an intellectual environment, one can imagine that one could believe that extraterrestrial microbes could exist, would never evolve and would never become anything other than microbes, but that is a point of view that may be hard to sustain against the knowledge that is rapidly accumulating under the umbrella of modern science.

Olson and Tobin argue that "Liturgical texts, insofar as they constitute the practical expressions of Orthodox theology, may be regarded as of an even greater authority than Patristic statements." On this basis, they point to the prayer for the Great Blessing of the Waters (commemorating the baptism of Jesus, which is celebrated by Orthodox Christians on the Feast of the Theophany, which falls on January 6). This prayer includes these words: "Great art Thou, O Lord, and marvelous are Thy works: no words suffice to sing the praise of Thy wonders. For Thou by Thine own will hast brought all things out of nothingness into being, by Thy power Thou dost hold together the Creation, and by Thy providence Thou doest govern the world. [...] All the spiritual powers tremble before Thee. The Sun sings Thy praises; the moon glorifies Thee; the stars supplicate before Thee."[7] Olson and Tobin interpret these words in this prayer to mean that not only the Earth but all of Creation is the "dominion of God." The conclusion naturally follows, they believe, that "everything within that universe—including any forms of life, microbial or otherwise—is a revelation of the existence and being of God."

Olson and Tobin point out that the biblical creation narrative in Genesis is geocentric and, as a result, so are the traditions of both the early Christian Church and the Orthodox Church, as these traditions have developed over many centuries; however, Olson and Tobin assert that this Earth-focused view is natural because "those who framed the creation narratives did so with a very restricted view of the form and structure of the universe." The Orthodox Church, though, has never demanded a literal interpretation of the creation story as a scientifically correct cosmology. In Orthodox thought, the Earth is special, even if it is not the only location in the universe with intelligent beings. The Earth may not be the physical center of the universe in Orthodox Christian cosmology, but, in the words of Olson and Tobin, "in the divine plan of salvation ... the Earth does indeed hold the central position, for it is the realm toward which God's love and redemptive work of Christ is directed."[8] They reiterate that "the non-central place of Earth within the galactic system does not contradict its central place within the wider sphere of Redemption and the priestly role of the Church in accomplishing the ultimate sanctification of all things even to the furthest reaches of the universe." Olson and Tobin conclude that the discovery of life beyond the Earth would reveal anew God's "infinity and omnipotence ... throughout the universe."

---

[7] From *The Festal Menaion, 1969* (Mother Mary & Archimandrite Kallistos Ware, Trans., p. 356). London, as quoted by Olson and Tobin in *Theology and Science*.

[8] Olson and Tobin, p. 432.

A handful of Orthodox theologians have taken a far more relaxed position, arguing that intelligent, spiritual beings could exist elsewhere and their existence would pose no threat to Orthodoxy. At the extreme end of the intellectual spectrum is Vladimir Lossky (1903–1958 C. E.), who appears willing to accept that Orthodox Christianity is a set of religious practices only for Christians on Earth. Other cultures elsewhere would have their own religious truths. Assuming they are religious at all, extraterrestrial beings would not necessarily need any knowledge about Christian revelation or about the Earthly Incarnation of the Son of God.

Lossky, a very influential twentieth-century Orthodox theologian, assumes that life likely exists elsewhere. In his 1944 book *The Mystical Theology of the Eastern Church*,[9] he suggests that "In the face of the vision of the universe in which the human race has gained since the period of the renaissance, in which Earth is represented as an atom lost in infinite space amid innumerable other worlds, there is no need for theology to change anything whatever in the narrative of Genesis; any more than it is its business to be concerned over the question of salvation of the inhabitants of Mars." Even though Martians, or other Others, may exist, he asserts that that information should not in any way affect Eastern Orthodox beliefs or practice. "Revelation," he writes, "remains for theology essentially geocentric, for it is addressed to men and confers upon them the truth as it is relative to their salvation under the conditions which belong to the reality of life on Earth."

More generally, Lossky writes about the relationship between Orthodoxy and science and asserts that Orthodox theology is infinitely flexible: "The Church always freely makes use of … the sciences for apologetic purposes, but she never has any cause to defend these relative and changing truths as she defends the unchangeable truth of her doctrines. … Modern cosmological theories cannot affect in any way the more fundamental truth which is revealed to the Church."[10] Furthermore, "Christian theology … is able to accommodate itself very easily to any scientific theory of the universe."[11] Thus, Lossky asserts, should life elsewhere be found, the Orthodox will have no problems accommodating that information without compromising their core values and beliefs.

The prominent twentieth-century Orthodox Christian priest and theologian John S. Romanides (1927–2001 C. E.), who was Professor of Dogmatic Theology at Holy Cross Theological School in Massachusetts, then Professor of Dogmatic Theology at the University of Thessalonika, Greece and finally Professor of Theology at Balamand Theological School, Lebanon, follows the lead of Lossky. In 1965 he penned an essay on extraterrestrial life that appeared in *The Boston Globe*. In his "All Planets the Same" piece he writes, "I can foresee no way in which the teachings of the Orthodox Christian tradition could be affected by the discovery of intelligent beings on another planet." He then notes that most of his colleagues felt that any time spent on the topic would be time wasted and certainly would not be considered a serious theological conversation. He seems to disagree with this assessment of irrelevance, noting that the existence of intelligent, extraterrestrial

---

[9] Lossky, V. (1944). *The mystical theology of the Eastern Church* (p. 105), translated from the Russian by the Fellowship of St. Alban and St. Sergius. London: James Clarke and Co: 1957, as quoted in Olson and Tobin.

[10] Lossky, p. 104, as quoted in Olson and Tobin.

[11] Lossky, p. 106, as quoted in Olson and Tobin.

life forms "would raise questions for traditional Roman Catholic and Protestant teachings regarding creation, the fall, man as the image of God, redemption and Biblical inerrancy." Such important theological issues, Romanides argues, are worthy of a serious, extended conversation.

Fortunately for Orthodox Christians, he points out, "Greek Christianity never had a fundamentalist or literalist understanding of Biblical inspiration and was never committed to the inerrancy of scripture in matters concerning the structure of the universe and life in it." No original sin. No fiery hell below. No starry heaven above. He suggests, "For the Orthodox discovery of intelligent life on another planet would raise the question of how far advanced these beings are in their love and preparation for divine glory. As on this planet, so on any other, the fact that one may have not as yet learned about the Lord of Glory of the Old and New Testament, does not mean that he is automatically condemned to hell, just as one who believes in Christ is not automatically destined to be involved in the eternal movement toward perfection. ... Thus the existence of intelligent life on another planet behind or way ahead of us in intellectual and spiritual attainment will change little in the traditional beliefs of Orthodox Christianity."[12]

Archbishop Chrysostomos of Etna offers a different perspective. Archbishop Chrysostomos's views emerge from a part of the Orthodox tradition that rejects the modern ecumenical movement and offers a fundamentalist approach to Orthodox Christianity. In taking this position, traditionalists like Archbishop Chrysostomos appear to be fairly far outside of mainstream Orthodoxy. They reject the idea that the many divisions of Christianity in the modern world are all co-equals in their various and varied understandings of Christianity. Presumably for Chrysostomos, non-Christian religious traditions must be even less acceptable. Writing in *Orthodox Tradition*, a journal published by the Center for Traditionalist Orthodox Studies, in his article "Alien Abductions and the Orthodox Christian," he demonstrates that a significant strain within Orthodox Christianity has not yet found a way to become reconciled with modern science. He then takes on the theologically unusual issue of alien abductions. Chrysostomos starts by offering a word of caution: alien abduction stories might not be true. He writes, "Whether or not these incidents are believable," but his words that follow strongly suggest he thinks many of these alien abduction events did occur. He continues, "as Orthodox Christians we believe that spiritual knowledge, not advanced technology, is the prime factor in the expansion and perfection of human consciousness. ... What, then, if they are not advanced beings from other planets, *are* these alien abductors? Ultimately, one cannot escape the conclusion that they are demons or phantoms created by demonic power. In the first place, they look like demons. ... If these aliens are not demons, how is it that beings so advanced that they can achieve space travel cannot prevent pain and scarring during routine physical examinations? It is not pain which the aliens cannot control, but their demonic passion for inflicting the same on mankind. ... Abductees are drawn away from the universal teachings of Orthodox Christianity and towards the demonic delusion that underlies modern New Age philosophies."[13]

---

[12] Romanides, J. S. (1965, April 8). All planets the same. *The Boston Globe*, p. 18. Retrieved from http://www.johnsanidopoulos.com/2009/11/fr-john-romanides-on-extraterrestrial.html

[13] Archbishop Chrysostomos of Etna. (1997). Alien abductions and the Orthodox Christian. *Orthodox Tradition, XIV*(1), 57.

Father Seraphim Rose (1934–1982 C. E.), who represents the same anti-modernist Orthodox Christian tradition as Archbishop Chrysostomos, offers a similar view in his 1975 book *Orthodoxy and the Religion of the Future*.[14] Rose asserts that extraterrestrial life does not exist, though alien demons do. Humankind, he writes, lacks "any reason to believe that there are 'highly evolved' beings on other planets." Nevertheless, Rose claims that many UFO sightings are real and that these UFO sightings and close encounters with aliens are not encounters with extraterrestrials from other worlds; rather, they are encounters with demons. Humankind, he tells his readers, "knows that there are indeed 'advanced intelligences' in the universe besides himself; these are of two kinds, and he strives to live so as to dwell with those who serve God (the angels) and avoid contact with the others who have rejected God and strive in their envy and malice to draw man into their misfortune (the demons)." Rose notes that history, in particular Orthodox literature, offers "many examples of demonic manifestations which fit precisely the UFO pattern. … The multifarious demonic deceptions of Orthodox literature have been adapted to the mythology of outer space, nothing more. … The 'visitors from space' theory is but one of the many pretexts they are using to gain acceptance for the idea that 'higher beings' are now to take charge of the destiny of mankind."

Father Thomas Kulp, who is identified in *Flying Saucer Review*, a journal that exists somewhere quite far beyond the even the fringes of mainstream science, as a priest of the Orthodox Church in Wisconsin, agrees. Writing in that journal in 2000, Kulp claims "There can no longer be any doubt that something significant is happening. On the whole, the reported incidents can neither be regarded as hoaxes, nor as some bizarre form of collective hallucination. Must we conclude, then, as many have, that extraterrestrial beings are visiting Earth?" No. But Kulp then draws a scientifically surprising, theologically challenging conclusion: "There is not a single UFO incident on record that cannot be explained as demonic deception or apparition. … the ultimate purpose of the UFO phenomenon is to help prepare the collective consciousness of the human race for the coming of Antichrist as foretold in the Bible."[15]

Thus, we find that the few mainstream Orthodox Christians who have made their views public on the matter of extraterrestrial life have low expectations for the idea that extraterrestrial life of any sort exists. They also all hold a high opinion of the centrality of humanity and the Earth in the story and purpose of the universe. But if extraterrestrial life is found, these Orthodox Christian leaders expect to easily accommodate that information, whatever it is. Another quiet voice within the Orthodox tradition, one that appears to be the voice of the few rather than the many, is very liberal and modernist, accepting the possibility of advanced extraterrestrial beings and asserting that their existence has no relevance for assuming the truth of Christian Revelation for Orthodox Christians on Earth. As for the Orthodox fundamentalists, they appear ready and eager to acknowledge that beings identifiable as aliens do exist, have already been discovered and have visited Earth multiple times. Those extraterrestrials, however, are fallen angels, otherwise known as demons, who are doing the bidding of Satan himself.

---

[14] Rose, S., Fr., (1975). *Orthodoxy and the religion of the future* (4th ed., pp. 77–114). Saint Herman of Alaska Brotherhood.

[15] Kulp, T. (2000). UFOs: A demonic conspiracy. *Flying Saucer Review, 45/3*.

10

The Church of England and the Anglican Communion

> *Christians have to ask themselves ... What can the cosmic significance possibly be of the localized, terrestrial event of the existence of the historical Jesus?*
>
> Arthur Peacocke (Peacocke, A. (2000). The challenge and stimulus of the epic of evolution to theology. In S. J. Dick (Ed.) *Many worlds: The new universe, extraterrestrial life, and the theological implications* (p. 103). Philadelphia, PA: Templeton Foundation Press.)

Those who have set forth their ideas on how extraterrestrials would fit into Anglican theology have offered every possible solution. Perhaps extraterrestrials will worship God but will not be Christians; perhaps they will worship the Christian Trinitarian God and know, somehow, about Jesus of Nazareth; or perhaps the necessity of knowing about the birth and death of Jesus in order to be a Christian, and the necessity of being a Christian in order to know God, is proof that humans on Earth are the only intelligent beings in the universe.

The Church of England, as well as the many Anglican and Episcopal Churches world-wide that are in communion with the Church of England, traces its roots back to the arrival of the first Christians in Britain in the third century C. E. In 1534 C. E., during the reign of King Henry VIII (1509–1547 C. E.) and after Pope Clement VII refused to annul Henry's marriage to Catherine of Aragon, the Roman Catholic Church in England broke with Rome and proclaimed itself an independent religious body, the Church of England. Theologically, the Church of England retains "a large amount of continuity with the Church of the Patristic and Medieval periods in terms of its use of the catholic creeds, its

D.A. Weintraub, *Religions and Extraterrestrial Life: How Will We Deal With It?*,
Springer Praxis Books, DOI 10.1007/978-3-319-05056-0_10,
© Springer International Publishing Switzerland 2014

pattern of ministry, its buildings and aspects of its liturgy, but which also [embodies] Protestant insights in its theology and in the overall shape of its liturgical practice." Thus, the Church of England describes its theology as both "catholic and reformed."[1]

The Anglican Communion, which includes about 85 million members worldwide, now embraces "three broad traditions, the Evangelical, the Catholic and the Liberal." The Evangelical tradition emphasizes more Protestant evangelical aspects, the Catholic tradition emphasizes more traditional catholic practices, and the Liberal tradition emphasizes "the use of reason in theological exploration."[2] Virtually all of what has been written by Anglicans regarding extraterrestrial life comes from scholars in the Catholic and Liberal traditions. Those from the Evangelical Anglican tradition likely have views more consistent with those discussed in the later chapter on evangelical and fundamentalist Christianity.

Oxford cosmologist E. A. Milne (1896–1950 C. E.) was awarded the prestigious Gold Medal of the Royal Astronomical Society in 1935. In his 1952 book *Modern Cosmology and the Christian Idea of God*, based on his 1950 Cadbury Lectures at the University of Birmingham, he took on the Thomas Paine issue of multiple incarnations and crucifixions. Milne asserts that the God of Christianity intervened uniquely in human history through the Incarnation: "Was this a unique event, or has it been reenacted on each of a countless number of planets? The Christian would recoil in horror from such a conclusion. We cannot imagine the Son of God suffering vicariously on each of a myriad of planets. The Christian would avoid this conclusion by the definite supposition that our planet is in fact unique. What then of the denizens of other planets, if the Incarnation occurred only on our own? We are in deep waters here, in a sea of great mysteries."[3]

Milne appears to accept that the only morally tolerable approach is to assume either that Christianity is a religion that is confined to our single world or that the unique Incarnation of Christ on Earth, somehow, offers salvation to those denizens of other planets. He does not argue that intelligent life is confined to the Earth; in fact, he suggests that we teach those other-worldly beings about the Incarnation of Jesus on Earth by sending them radio signals. Such communications, he believes, would allow the single, historical Incarnation on Earth to affect the entire universe. (He ignores the fact that radio waves travel at the speed of light and not at infinite speeds, putting most of the universe beyond the spatial and temporal reach of his missionary dreams.) While offering this suggestion to theologians, as a cosmologist he does not feel compelled to solve the theological problem he and Thomas Paine recognize as critical to understanding Christianity's role in a universe in which humans on Earth are not the only sentient beings; however, unlike Thomas Paine, he does not lampoon the issue. Rather, from the perspective of a religiously devout Anglican who was also a world-renowned astrophysicist, he set before twentieth-century theologians his learned opinion that the likelihood that extraterrestrial life does exist in the modern universe is extremely high. He then demanded that Anglican theologians offer their followers a theologically satisfactory explanation as to how Christianity works in a universe with both humans and aliens.

---

[1] From http://www.churchofengland.org/about-us/history/detailed-history.aspx
[2] Ibid.
[3] Milne, E. A. (1952). *Modern cosmology and the Christian idea of God*. Oxford: Clarendon Press.

Milne provoked a pointed response from Anglican theologian E. L. Mascall (1905–1993 C. E.), who served as Professor of Historical Theology at King's College London. Mascall notes in his 1956 book *Christian Theology and Natural Science* that simply knowing about the Incarnation of the Son of God on Earth is not sufficient to bring about salvation for God's extraterrestrial creatures. In Mascall's theology, the facts that "the Son of God became man" and "lived and died as a man" are critical to the salvation of humankind, and only to humankind.[4] If Martians exist, Mascall writes, "redeemed Martians will be all one Martian in the Word-made-Martian; and this would seem to require an incarnation of the Word in Martian nature as its foundation." That is, if there are other fallen beings, God will and must become incarnate in the bodies, in the flesh, of those beings in order to offer salvation to those beings. While this view would seem to limit the power of God, Mascall does not think in those terms. He manages to keep all possibilities on the table by then suggesting, "it may or may not be such as to require an incarnation of the Son of God in Martian nature. ... God *might* have worked redemption in some other way. ... God may or may not have some other way of restoring fellowship with himself." Mascall covers all angles but takes no position—or perhaps takes every position—theologically. For Mascall, extraterrestrial life may or may not exist; if other-worldly creatures exist, they may or not be fallen; and if they are fallen, the Incarnation of the Son of God in the bodies of those creatures is possible but may or may not be God's only method for offering them salvation. One could characterize Mascall's theological views as either extremely open-minded or without any solid foundation, but one cannot accuse him of being doctrinally close-minded.

Oxford Don C. S. Lewis (1898–1963 C. E.) became a communicant in the Church of England in 1931. He then became one of the most popular and influential Christian apologetical writers of the twentieth century. Lewis presents an interesting mix of open-mindedness toward extraterrestrial life with Christian apologetics in his 'Religion and Rocketry' essay, penned in 1960.[5] Lewis warns that the discovery of animal life on another planet would be "held at first to have the most wide-reaching theological and philosophical consequences [and would be] seized by unbelievers as the basis for a new attack on Christianity. ... The supposed threat is clearly directed against the doctrine of the Incarnation."

The threat, he explains, would be about the apparent uniqueness of humankind: "Why for us men more than for others?... how can we, without absurd arrogance, believe ourselves to have been uniquely favored." After restating the Thomas Paine argument, Lewis attempts to counter it. He first places pre-conditions on the conversation about the salvation of extraterrestrials, which he assumes is the conversation we must have. There is little point, he argues, in having a perfectly 'what if' argument. We should only entertain this discussion seriously, Lewis says, if and when we identify extraterrestrial animals and can ascertain that they are both rational and "like us, fallen." Lewis takes the position that we should not think that humanity is uniquely favored. Rather we should recognize that

---

[4] Mascall, E. L. (1956). *Christian theology and natural science* (pp. 36–46). London: Longmans, Green and Co.

[5] Lewis, C. S. (1960). Religion and rocketry. In *The world's last night, and other essays* (pp. 84–91). New York: Harcourt, Brace and Company.

humanity has been redeemed because humanity is uniquely and particularly demented and depraved. "Perhaps," he writes, "of all races only *we* fell."

At this point, Lewis has accepted as likely the possibility that Paine is right, that humanity is the singularly worst behaving, most morally and ethically flawed species in the universe, the only species in need of redemption. If this is the case and Paine is right, then Christianity is a unique and special religion needed only by humans descended from Adam, i.e., earthlings. Lewis, however, is not willing to completely accept this, though this scenario is, for him, a real possibility.

Lewis does open his theology up to the possibility that extraterrestrials could be sinful and could need to be redeemed. In such a case, God might make the salvation history on Earth known and available to Others. Lewis writes that it is possible that only through "the birth at Bethlehem, the cross on Calvary and the empty tomb—a fallen race could be rescued. There may be a necessity for this, insurmountable, rooted in the very nature of God… But we don't know."

Furthermore, according to Lewis humankind may possibly occupy "a pivotal position" and that redemption of other species works through and even depends on us. Despite contemplating that man might occupy "a pivotal position" in the universe, he claims, perhaps disingenuously, "such a position need not imply any superiority in us or any favouritism in God." At this point, Lewis' views seem to be a have-your-cake-and-eat-it-too approach: we are not special or superior, but we are the pivot point for the salvation of the entire universe. An adjudicator likely would find these ideas in conflict; if we are the pivot point for the salvation of the entire universe, we necessarily are also very special creatures in the eyes of God.

Lewis ultimately appears to take the position that Christianity is or should be the religion of the entire universe and that "the mere existence of extraterrestrials would not raise a problem." We just need to figure out which of those creatures are fallen and deserve redemption. Those creatures without sin will receive redemption without paying any attention to the cross on Calvary while those creatures in need of redemption will somehow receive it because of the events that took place on Earth 2,000 years ago. In fact, once we discover aliens, he expects that "we shall perhaps send missionaries," though he thinks that future missionary work will go as badly as the missionary work conducted in Africa, Asia and the Americas by white Europeans over the last few centuries. Should we meet extraterrestrials, humans will "commit all the crimes we have already committed against creatures certainly human but differing from us in features and pigmentation." Though condemning the missionary work of European Christians as carried out in the last millennium, Lewis appears to believe that Christians are fated, upon discovering extraterrestrials, to repeat the sins of their collective past, as they will feel compelled to proselytize extraterrestrials.

In his 2005 book *Creation and Double Chaos*, Sjoerd Bonting expresses his disappointment that "few contemporary theologians show much interest in the matter of possible extraterrestrial life."[6] As a result, he asserts that "the near absence of sound theological reflection on this topic in Bible and tradition and by prominent contemporary theologians necessitates some pioneer work in formulating a theology of extraterrestrial life," which

---

[6] Bonting, S. L. (2005). Are we alone? Theological implications of possible extraterrestrial life. In *Creation and double chaos: Science and theology in discussion*. Minneapolis: Fortress Press.

he then goes on to present. Such an exotheology, of course, only makes sense if one believes that extraterrestrial life might exist, and Bonting firmly does believe this, though he leans toward the side of the conversation that believes worlds with life are likely rare: "This may leave few fitting candidates, but … we cannot rule out that there are some."

Bonting's training in biochemistry leads him to think that "such life will be based, like all earthly life, on carbon chemistry," that extraterrestrials "would not be radically different from *Homo sapiens* in physiology and biochemistry," would "have brains and neuronal systems resembling ours" and "would have similar thought processes." They will be mortal and the product of biological evolution on their home planets. Bonting then offers a very human-centered and Jesus-centered exotheology. In Bonting's theology, extraterrestrials are just like humans in that they somehow have become separated from God (i.e., they are sinful beings); also like humans, they seek to become reconciled with God. Sin, to Bonting, is not something that occurred and entered human nature in a historical, physical Eden on Earth. In agreement with Tielhard de Chardin, Bonting argues that sin is some condition that simply exists throughout the universe, and so extraterrestrials will be sinners "just as much in need of salvation as we are."

The God in which Bonting believes "is the universal, omnipresent God of all peoples, Jews and Gentiles, and thus also of any extraterrestrials." Through Bonting's logic, the universality of his God and of sin means that "the creative work of the Father, the saving work of Christ, and the communicative action of the Holy Spirit will apply just as much to any creature on another planet as they do to us." According to Bonting, God will provide saving grace to humans as well as to extraterrestrial beings through Christ's incarnation and resurrection on Earth. Extraterrestrials will "participate in the reconciliation brought about by Christ's incarnation, death and resurrection two thousand years ago in Palestine, without necessarily requiring a repetition of these events on their planet. And as God has made the message of Christ's saving work heard in all times and in all corners of our planet, so he will also bring it in an appropriate way to any of his creatures on another planet."[7]

Bonting parries Thomas Paine's thrusts at Christianity through the assertion of "the uniqueness of Christ's incarnation" and the "cosmic significance and lasting validity" of that singular event. The one unique incarnation and resurrection in Palestine "brings salvation to us, who live two thousand years later in other parts of the planet, yes, to all humans who ever lived on Earth at any time and at any place. And not only to humans, but to the whole creation 'that has been groaning in labor pains until now' and 'waits with eager longing' for its final liberation, as Paul says (Romans 8:19, 22). Why not then to creatures on another planet? There is no need to assume multiple incarnations!" Somehow, knowledge of Christ will be made known by God to all species in all places in the universe at all times past and present—"The one God of the universe will have made himself known to them, as God has to us." In advocating this position, Bonting's theological training has trumped his professional training in the natural sciences, since the laws of physics would otherwise limit how far information from Earth can travel through the universe in two thousand years and certainly would seem to preclude sending information into the past.

---

[7] Bonting, S. L. (2003, September). Theological implications of possible extraterrestrial life. *Zygon, 38*(3).

For Bonting, the 'message of Christ' is universal—"I would even claim that salvation and reconciliation will come to them at the same time as to us, namely, when Christ at his triumphant return will definitely banish the remaining element of chaos from the universe." He believes that the 'message of Christ' is also one that all intelligent and necessarily sinful beings beyond the Earth must hear. He does not, however, concern himself with how God will accomplish the task of sharing this message with Others. Bonting does not address, or perhaps even recognize, the philosophical problems associated with this solution, in which the entire universe is Christian but humankind occupies the central place in God's creation.

Lewis's and Bonting's Anglican exotheology emphasizes the need for redemption and concerns itself with the fact of human (and perhaps Other) sinfulness but not at all with the Roman Catholic issue of original sin. For Lewis and Bonting, the redemptive act of the Son of God was a unique event that occurred on Earth, but that redemptive act, which defines Christianity, somehow will become known throughout the universe, for all time past and future, for the benefit of all fallen species everywhere.

The Reverend Brian Hebblethwaite (b. 1939), the Canon Theologian of Leicester Cathedral, England, from 1983 until 2001, agrees with one part of Lewis's position, but then draws a very different conclusion. He argues in agreement with Lewis, in his essay "The Impossibility of Multiple Incarnations" penned in 2001, that multiple incarnations are impossible, but he also believes that the incarnation of Christ is a necessary prerequisite for salvation. He then marches to a different drummer when he concludes that since extraterrestrial species cannot benefit from their own incarnations, extraterrestrial life cannot exist.[8]

The Reverend Canon Arthur Peacocke (1924–2006 C. E.), who earned a D. Phil. at Oxford in 1948 in physical biochemistry and was ordained as a priest in the Church of England in 1971, held a very expansive view of the likelihood that extraterrestrial life exists. Peacocke writes in his 1993 book *Theology for a Scientific Age* that "In those parts of the universe where the temperature is low enough for molecules to exist in sufficient proximity to interact, there is a tendency for more and more complex molecular systems to come into existence," a process he identified as a propensity for increased complexity.[9] That propensity for increased complexity, he argues, has a high probability of leading to something we would call alive and for those living things to develop intelligence. Seven years later, in his essay "The Challenge and Stimulus of the Epic of Evolution to Theology," he writes "If the chemical conditions were right on a planet of about the same age as the Earth, moving round a planet of about the age of our Sun, then it is probable that living forms of matter would have appeared on it; and with a lower nonzero probability, that intelligent creatures would have emerged by operation of natural selection."[10]

Of course, Peacocke appears to have hedged his bet, since he begins with "if." We need the right chemical conditions and the right kind of planet in the right location. Maybe

[8] Hebblethwaite, B. (2001). The impossibility of multiple incarnations. *Theology, 104*(821), 323–334.

[9] Peacocke, A. (1993). *Theology for a scientific age* (pp. 65–66). Minneapolis: Fortress Press.

[10] Peacocke, A. (2000). The challenge and stimulus of the epic of evolution to theology. In S. J. Dick (Ed.) *Many worlds: The new universe, extraterrestrial life, and the theological implications*. Philadelphia, PA: Templeton Foundation Press.

such conditions are rare? But no. Peacocke runs the numbers and concludes that with at least 100 billion billion (10^{20}) stars in the universe, "the tiniest probability of extraterrestrial life still leads to a finite probability of its existence on a planet other than the Earth at some time." Given this near absolute certainly that extraterrestrial beings exist, did exist, or will exist, Peacocke asks the question that theologians have tried to avoid asking: "Christians have to ask themselves … What can the cosmic significance possibly be of the localized, terrestrial event of the existence of the historical Jesus? … Would ET, Alpha-Arcturians, Martians, et al., need an incarnation and all it is supposed to accomplish? … Only a contemporary theology that can cope convincingly with such questions can hope to be credible today."

Peacocke offers such a theology, one in which Christianity is not only unique to Earth but is "unique to Christians." For Peacocke, "the whole Christ event" shows God acting in history to show "what is possible for humanity" and "what God intends for all human beings, … of becoming united with God." Such an intervention of God in the history of life on a planet could "in principle, be manifest both in other human beings and indeed also on other planets, in any sentient, self-conscious, *nonhuman* persons." Peacocke offers no worries or concerns about sin and redemption or about how knowledge of the resurrection might be passed around the universe. To him, all of these issues are restricted to terrestrial Christianity. Elsewhere in the universe, wherever there are beings capable of relating to God, God will intervene in the history of those beings in a species-appropriate way.

No single doctrinal authority speaks for the greater Anglican community. Those who have positions of leadership within this faith and who have given voice to their opinions on the subject have offered every possible solution to the Thomas Paine problem. While Anglicanism has not found a single, broadly acceptable solution to the problems associated with extraterrestrial life and Anglican theology, Anglicans have been actively engaged with the issues and appear ready to deal with whatever astronomers discover, though some proposed solutions will necessarily fail while others will prove more robust.

11

Mainline Protestant Christianity

> *Man cannot claim to occupy the only possible place*
> *for incarnation.*
>
> Paul Tillich

Protestantism arose in the early sixteenth century as a reaction against certain practices of the Roman Catholic Church. In particular some deemed the selling of indulgences a questionable practice. The German priest and theologian Martin Luther argued that human actions of any kind have no role whatsoever in divine grace. Consequently, he argued, God's grace could not be purchased via an indulgence. The initial schism that started with Luther in Wittenberg quickly fragmented as the Lutherans (Germany), Calvinists (Switzerland), Anglicans (England), Huguenots (France), Anabaptists (Switzerland), Presbyterians (Scotland), Baptists (The Netherlands) and Congregationalists (England), among other groups, navigated different paths in defining Reformation era Christianity. New denominations have continued to form with regularity since the early decades of the Reformation.

A fundamental concept that underpins all of Protestantism and that emerged from Martin Luther's writings is *sola scriptura* ("by scripture alone"). According to the doctrine of *sola scriptura*, Holy Scripture contains all the information necessary for an individual to achieve salvation. In addition, according to the early Protestant reformers, this knowledge is made clear in scripture to both the educated and the uneducated, alike. In contrast to Roman Catholicism, no intermediaries—no popes, bishops, priests or church councils—are needed to interpret these words for a believing Christian.

A second of Martin Luther's ideas that is common to all Protestant denominations came to be known as the *priesthood of all believers*. This idea permits and even encourages each and every Protestant believer to interpret scripture independently. While many

D.A. Weintraub, *Religions and Extraterrestrial Life: How Will We Deal With It?*,
Springer Praxis Books, DOI 10.1007/978-3-319-05056-0_11,
© Springer International Publishing Switzerland 2014

believers are happy to be told what to believe by leaders of their respective denominations, whether these are the interpretations of scripture according to Martin Luther, John Calvin, Billy Graham or Joel Osteen, Protestantism in general allows great license for each individual to interpret scripture by himself or herself and in his or her own way. Since all Protestants may interpret scripture, whether they are well or poorly educated, the combination of *sola scriptura* and the *priesthood of all believers* can, though it does not have to, lead to a de-emphasis on the use of logic, reason, science and history in understanding and interpreting scripture. As a result, the many different possible Protestant approaches for understanding and interpreting scripture lead in many, often opposite, directions in the context of religious beliefs regarding extraterrestrial life.

Early Reformation movements tended to be literalist in interpreting scripture. The emphasis on the Word of God rather than on the interpretation of those words by councils of Catholic bishops or by great scholars led to a great diversity of ideas among the first Protestant sects, but also to very strict beliefs among adherents within each group. Later, during the Enlightenment, influential Protestant intellectuals began to argue against biblical inerrancy and literalism; as a result, some Protestant denominations became more open to the use of science and philosophy in their efforts to understand Christian doctrine.

In the nineteenth and early twentieth centuries, many fundamentalist and evangelical churches (e.g., Seventh-day Adventists, Jehovah's Witnesses) appeared and grew their membership by embracing an attitude that the established (mainline) Protestant churches were not upholding the inerrancy of biblical teachings, especially in increasingly "progressive" or "liberal" stances regarding evolution, the use of alcohol and the rights of women. These same attitudes, in the late twentieth and early twenty-first centuries, are manifest in the opposition by many fundamentalist churches and individuals to gay rights and evolution and in their reluctance to use scientifically-derived data to interpret the world around us.

In the first decades of the twenty-first century, the two foundational elements of Protestantism have allowed the mainline Protestant denominations in America (most Methodists, Presbyterians, Congregationalists, Lutherans and northern Baptists) to become increasingly open to change on social issues, like the ordination of women and of gay and lesbian pastors. These same foundational elements, however, have also led over the last century to the birth and growth of even more fundamentalist and evangelical churches, whose members and leadership interpret and understand parts of scripture very differently from their mainline counterparts.

For all Protestants, whether mainline or evangelical, the idea of extraterrestrial life is problematic because this concept is not directly addressed in the Christian scriptures. Consequently, if science does discover extraterrestrial life, this knowledge will emerge from completely outside of scripture.

Paul Tillich (1886–1965 C. E.), an ordained Lutheran minister, was one of the most influential Protestant theologians of the twentieth century. Tillich agreed that theologians must wrestle with the concept of extraterrestrial life, writing in his 1950s multi-volume *Systematic Theology*, "A question arises which has been carefully avoided by many traditional theologians, even though it is consciously or unconsciously alive for most contemporary people. It is the problem of how to understand the meaning of the symbol 'Christ' in the light of the immensity of the universe, the heliocentric system of planets, the infinitely small part of the universe which man and his history constitute, and the possibility of other 'worlds' in which divine self-manifestations may appear and be received."[1]

For Tillich, the need for salvation is universal and the 'saving power' of God (the divine influence that enables living creatures to earn salvation) must be everywhere. "The independence of everything with everything else in the totality of being," he writes, "includes a participation of nature in history and demands a participation of the universe in salvation. Therefore, if there are non-human 'worlds' ... such worlds cannot be without the operation of saving power within them. ... The manifestation of saving power in one place implies that saving power is operating in all places."

According to Tillich, however, while God's saving power may be universal Christianity is localized to those who know about Jesus. Christianity, he writes, "is what it is through the affirmation that Jesus of Nazareth ... is the actual Christ." If salvation occurs in other ways, even if God's methods involve incarnation, those creatures will be redeemed, but in Tillich's view they would not be Christians. "Our basic answer leaves the universe open for possible divine manifestations in other areas or periods of being. Such possibilities cannot be denied. But they cannot be proved or disproved. Incarnation is unique for the special group in which it happens, but it is not unique in the sense that other singular incarnations for other unique worlds are excluded. ... Man cannot claim to occupy the only possible place for incarnation. ... in the appearance of Christ he actualizes this love for historical man alone."

Tillich's God is a truly universal God. On the other hand, in his view humanity's struggle with sin and our need for redemption and salvation appear to be part of God's plan for human life. They are not of necessity God's plan for intelligent life elsewhere in the universe. Tillich suggests that we need to interpret Christianity as the way in which God's grace was given to humanity. For each species of Others, if an incarnation is part of God's plan then that event will occur in a way appropriate for that species. In Tillich's views, the Incarnation of Christ on Earth followed by the crucifixion and resurrection was a unique series of events that provides the structure for the salvation of humanity; the sequence of events surrounding the life of Jesus allows the possibility for humans to be Christians. Other life forms, whether they do or do not experience their own incarnation event, will not be Christians. Tillich does not focus on original sin, the Trinity, redemption, or the need for God's mercy for extraterrestrials. From his perspective, some or all of these pieces of the religious puzzle may be ours and ours alone with which to wrestle. Tillich's approach represents the extreme, liberal end of Protestantism in regard to the question of extraterrestrial life.

The school of process theology, which in the early twentieth century emerged and built on the theological ideas of Alfred North Whitehead (1861–1947 C. E.), offers a similar view to that put forward by Tillich. From the perspective of process theology, Christianity is a terrestrial religion meant for humans alone, though all of God's creatures require some form of salvation. Process theologian Lewis S. Ford (b. 1933), Emeritus Professor of Philosophy at Old Dominion University and the founding editor of the journal *Process Studies*, offers a skeptical view of the uniqueness of the terrestrial incarnation in his 1978 book *The Lure of God*: "Are we then to conclude that God's only son became

---

[1] Tillich, P. (1951–1963). *Systematic theology* (vol. II, pp. 95–97). Chicago: University of Chicago Press.

uniquely incarnate once and for all on the third planet of a rather ordinary star of a thoroughly undistinguished galaxy?"[2] In his essay "Theological Reflections on Extra-terrestrial Life" written a decade earlier, Ford is willing to accept that the concept of salvation may apply to extraterrestrials—"Salvation is not just limited to men but applies to all intelligent beings wherever they dwell"[3]—but he believes God has the power and therefore the flexibility to offer salvation to Others in ways other than the Christ. He argues that we don't know God's purpose for all places and all times. Instead, we can only think about God having a purpose for God's actions.

With reference to God's purpose for humanity, the answer is clear to Ford: "We understand by the Logos or divine creative Word the sum and totality of all God's specific creative purposes for all creatures," he writes in *The Lure of God*. "This creative purpose is hardly invariant in its specific manifestations: what God says depends upon the particular situation confronting that individual in his own world. … God's dynamic word knows no single form. … The Logos, then, refers to the totality of God's creative aims. We may distinguish this from the Christ, which signifies one specific divine creative purpose addressed to the human situation. … The Word appropriate to our condition becomes incarnate by becoming fully actualized in the words, deeds, and suffering of Jesus."[4]

Physicist and process theologian Ian Barbour (1923–2013 C. E.), who spent his career on the faculty at Carleton College, draws the same conclusion. In his 1997 book *Religion and Science*,[5] Barbour makes clear that he adheres to the cosmological principle, that the laws of physics at work in our part of the universe are the same laws of physics that operate in all other parts of the universe. In Barbour's words, "The cosmos is all of a piece. … Humanity is the most advanced form of life we know, but it is fully a part of a wider process in space and time." Given that the physics and chemistry of the universe led to the emergence of life in our part of the universe, Barbour sees no reason why life could not have emerged in other stellar systems. Given that extraterrestrial life is possible, if not likely, "the possibility of beings superior to us," he writes, "living in more advanced civilizations, is a further warning against anthropocentrism. It also calls into question exclusive claims concerning God's revelation in Christ." Assuming, as Barbour does, that the Word of God "was creating throughout the cosmos," then the Word of God "will also have revealed itself as the power of redemption at other points in space and time, in ways appropriate to the forms of life existing there."

For Tillich, Ford and Barbour, all intelligent beings need to be redeemed and merit salvation from the one God of the universe. Why salvation is an issue for all such creatures is unexplained, but both the heavy burden of original sin arising from the Garden of Eden alone and the necessity of redemption uniquely through the sacrifice of Jesus of Nazareth, so ridiculed by Thomas Paine, are absent from these theological approaches. For these theologians, the God of Christianity is a universal God and sin is a universal condition of rational beings, but for them Christianity is a terrestrial and not a universal religion.

[2] Ford, L. S. (1978). *The lure of God* (p. 63). Philadelphia: Fortress Press.

[3] Ford, L. S. (1968, Fall). Theological reflections on extra-terrestrial life. *Raymond Review, 3*(1), 2, as quoted in Peters, T. (2003). *Science, theology, and ethics* (p. 128). Ashgate science and religion series.

[4] Ford, L. S. (1978). *The lure of God* (pp. 63–64). Philadelphia: Fortress Press.

[5] Barbour, I. G. (1997). *Religion and science* (p. 215). San Francisco: Harper.

Ted Peters (b. 1941), a Professor of Systematic Theology at Pacific Lutheran Theological Seminary in Berkeley, California, has been thinking about the theological issues associated with extraterrestrial life for half a century. He concludes, in his 1975 essay "Exo-Theology," that "although there are partial grounds for thinking the Christian faith is so Earth centrist that it could be severely upset by confirmation of the existence of ETI, an assessment of the overall historical and contemporary strength of Christian theology indicates no insurmountable weakness." Peters recognizes that Christian fundamentalism may be the exception to this rule, but avers that "it would be a mistake to take the fundamentalist fright as representative of Christianity as a whole." While Peters himself makes no attempt to solve the theological problems associated with the existence of extraterrestrials, he does address some of them. As for "this debate over the need for multiple incarnations, we need to keep one item in mind. Even though there are slight differences of opinion regarding the relationship between ETI and the historical event of redemption on Earth, what is important is the common assumption that possible ETI belong within the realm of God's creation."[6]

German systematic theologian Wolfhart Pannenberg (b. 1928), Professor of Theology Emeritus at the University of Munich, has written that "God's action, then, is seen to be a single act that embraces the whole cosmic process ... and that thus leaves room for a plurality of creatures." Though Pannenberg's notion of 'creatures' is clearly targeted at terrestrial beings, as an aside he wonders about creatures elsewhere: "But are we humans really the goal of creation? ... Might not there be, in other galaxies and solar systems, planets on which life might arise and on which there might be intelligent beings?" Having accepted the notion that life beyond Earth could exist, he dismisses such life as non-threatening to Christian teaching, though he ignores all the theological details: "[T]he as yet problematic and vague possibility of their existence in no way affects the credibility of Christian teaching." As for the universality of Christianity, he puts forward the idea that "the turning of the Father to each of his creatures ... is always mediated through the Son." As such, Pannenberg asserts a universal significance for the Christian Trinitarian concept of God. Beyond such bland statements, Pannenberg has nothing to say concerning the theological issues related to extraterrestrials.[7]

Similarly, German Reformed theologian Jürgen Moltmann (b. 1926), Professor Emeritus of Systematic Theology at the University of Tübingen, accepts and even embraces the concept of extraterrestrial life. Moltmann also argues for the cosmic significance of Christ, but not for a cosmic Christianity. In wondering about the uniqueness of the universe, Moltmann asserts that the evolution of life as it occurred on Earth happened only once in the universe; however, he writes, "perhaps on other planets we shall find comparable evolutions of life in stages different from those on earth."[8] In Moltmann's cosmotheology, the many other worlds, and presumably the living beings on those many other worlds at whatever their evolutionary stages, do not belong "to many gods or powers. This [the entire universe] is *the one* creation of *the one* God. ... The fellowship of all created

---

[6] Peters, T. (1975). Exo-theology. In *The Gods have landed: New religions from other worlds* (pp. 187–206). Albany: State University of New York Press.

[7] Pannenberg, W. (1994). Systematic Theology, v. 2 (G. W. Bromiley, Trans., pp. 21, 34–35, 74, 76). Grand Rapids, MI: William B. Erdmans Publishing.

[8] Moltmann, J. (2003). *Science and wisdom* (M. Kohl, Trans., p. 83). London: SCM Press.

beings goes ahead of their differentiations and the specific forms given to them."[9] Moltmann further envisions a "cosmic christology" in which Christ died "so as to reconcile everything in heaven and on earth … and to bring peace to the whole creation. … The transition of Christ," he writes, "has more than merely historical significance. It has cosmic meaning too. Through this transition resurrection has become the universal 'law' of creation, not merely for human beings, but for animals, plants, stones and all cosmic life systems as well."[10] No rational explanation for how this might happen is necessary for Moltmann; the details of religious worship, faith and practice, from worries about sinfulness to concerns about salvation, are irrelevant to him as to how this cosmic reconciliation will occur. Moltmann makes his vision very clear: "We don't want to spread Western civilization. We want to invite people in all civilization to the new creation of all things. We don't want to expand the sphere of influence of the church. … We shouldn't try to turn everyone into Lutherans or Baptists, or found Roman Catholic congregations everywhere. But wherever we proclaim God's kingdom, God's people gather together, just of themselves, and will have their own experiences and develop their own forms of belief and worship. The new creation is as rainbow-hued and diversified as creation at the beginning." He concludes by noting, "The kingdom of God isn't there for the sake of the church. The church is there for the sake of the kingdom."[11]

In his 2000 essay "Extraterrestrial Life and the Cosmic Christ as Prototype," Martin Thomson, a Minister in the presbyterian Church of Scotland, examines the question as to what being made "in God's image" means for Christians when they are confronted with the possibility of intelligent extraterrestrial beings. Human beings are "God's representatives on earth," he writes, but the concept of image need not have anything to do with physical characteristics. "There is nothing to preclude the possibility of that image being manifest in alien life." The universe, in fact, may be "teeming with life bearing his image," which in Thomson's opinion would only serve to further glorify Christ. With this in mind, Thomson puts forward a cosmic Christology in which "there is no domain over which Christ does not hold sway." Thomson then echoes the views of John Jefferson Davis, a Presbyterian pastor and Professor of Systematic Theology and Christian Ethics at Gordon-Conwell Theological Seminary (Massachusetts), saying "the redemptive effects of the atoning sacrifice of Christ are not limited to humanity, but extend in some way to the entire created order." Thomson reminds us that, as normally understood, the Westminster Confession of Faith—a Reformed confession of faith, originally drawn up by Puritan clergymen for the Church of England during a series of meetings in Westminster Abbey in London in 1646, that remains a standard for the Church of Scotland and is influential within Presbyterian churches worldwide—does not limit the effectiveness of the incarnation of Christ by time. That is, the laws of physics do not apply. With this in mind, Thomson has no problem in also applying the idea of the broad effectiveness of the incarnation of Christ to space: the efficacy of the message of Christ is not limited by the vastness of space

---

[9] Moltmann, J. (1994). *Jesus Christ for today's world* (M. Kohl, Trans., p. 95). Minneapolis: Fortress Press.

[10] Moltmann, J. (1993). *The way of Jesus Christ* (M. Kohl, Trans., pp. 255–258). Minneapolis: Fortress Press.

[11] Moltmann, J. (1994). *Jesus Christ for today's world* (M. Kohl, Trans., pp. 146–147). Minneapolis: Fortress Press.

on a cosmic scale.[12] Thomson's ideas bear some similarities to the more refined ideas expressed by Moltmann, in which the resurrection of Christ is believed to have significance for all of time and all of space, even though the specific concepts and precepts of Christianity as practiced by members of the Church of Scotland may be irrelevant on other worlds.

The United Church of Christ (UCC) is unusual, in that it is an organization that has taken a formal position on the issue of extraterrestrial life. That position can be summarized in only a few words: They probably exist so we should start talking about them. Formed in 1957 when followers of four Protestant traditions—the Congregational Churches of the English Reformation with Puritan New England roots in America; the Christian Church with American frontier beginnings; the Evangelical Synod of North America; and the Reformed Church in the United States—united to form an umbrella organization that now includes more than 5,000 Protestant churches with more than one million members. The UCC has no centralized authority and no hierarchy that can impose doctrinal uniformity on its members or member churches; however, the UCC does have leaders and spokespersons who offer certain points of view that they encourage member churches to follow.

The Reverend John H. Thomas, General Minister and President of the United Church of Christ from 1999 to 2009, wrote a pastoral letter in 2008 entitled *A New Voice Arising: A Pastoral Letter on Faith Engaging Science and Technology,*[13] in which he asks "Are we alone? … There are good reasons to think that life, even intelligent life, exists throughout the universe. … Gone is the old view of a small, static universe, with fixed species dwelling on a fixed earth." In his letter, Reverend Thomas does not identify any theological issues UCC members will have to address regarding the existence of extraterrestrials, but he does offer the opinion that "Our faith has nothing to do with clinging to ancient misconceptions. … we are aware that we will have to expand and modify our responses whenever we are met by new concepts and unexpected possibilities." Via this letter, Reverend Thomas encouraged all UCC churches to designate one Sunday each year to engage in dialogue about how science and technology affect the faith of members. Presumably, some of those conversations will focus on extraterrestrials and exotheology.

Reverend Thomas's approach, along with the exotheology of Tillich, which is generally echoed by Ford, Barbour, Peters, Pannenberg, Moltmann and Thomson, appears to offer a viable, consensus approach for most mainline Protestants and their independent churches and denominational organizations. From this theological perspective, Christians from many Protestant traditions are unlikely to assume that Christian ideas about incarnation and redemption have any place beyond the Earth on planets that house advanced life forms, but they are likely to assume that sin and the need for God's grace, in whatever form it might be bestowed, is universal.

---

[12] Thomson, M. (2000, Autumn). Extraterrestrial life and the cosmic Christ as prototype. *Scottish Bulletin of Evangelical Theology, 18.2,* 160–178.

[13] Thomas, J. H. (2008). *A new voice arising: A pastoral letter on faith engaging science and technology.* Retrieved from http://www.ucc.org/not-mutually-exclusive/

12

Evangelical and Fundamentalist Christianity

Physical life was specially created, and Earth was created uniquely to support that life.

Henry Morris

Arthur C. Clarke, in his short-story "The Star" written in 1955, elucidates with the kind of prescience typical of his work the inevitable conflict between biblical literalism and scientifically-derived knowledge as it might relate to the discovery of life on distant worlds. In "The Star," a Jesuit priest trained also as an astrophysicist travels thousands of light years across our galaxy to explore the Phoenix Nebula. The Phoenix Nebula is a giant, rapidly expanding cloud of hot gas left behind when a star ran out of nuclear fuel and perished as a supernova. When a dying star that becomes a supernova explodes, it briefly shines as brightly as ten billion Suns. If that supernova exploded somewhere in the Milky Way galaxy, it would initially appear brighter than the brightest star in the Earth's nighttime sky. Then, after only a few days, it would begin to dim. After a few months, it would have faded away until it had become so faint that it would forever after elude detection by the unaided human eye.

When Clarke's priest reaches the Phoenix Nebula and begins exploring, he discovers a beacon on a Pluto-like planet located on the edge of what was once a planetary system. On that remote planet, he finds a monolith that marks the location of a Vault. The Vault contains messages and pictures left there millennia ago by the last living members of an ancient civilization. Those who built the Vault knew that they were about to perish in the shock wave that was about to engulf them in the form of million-degree gas boiling outwards in the expanding cloud spit out by the supernova explosion. After collecting all the information he could about the nebula and about the "disturbingly human" people who perished in the cataclysm, about their languages and arts and about "the warmth and beauty of a civilization that in many ways must have been superior to our own," the priest

D.A. Weintraub, *Religions and Extraterrestrial Life: How Will We Deal With It?*,
Springer Praxis Books, DOI 10.1007/978-3-319-05056-0_12,
© Springer International Publishing Switzerland 2014

calculates exactly when the former Sun-like star exploded. He realizes, in a discovery that challenges his faith, exactly when the light from this incredibly bright, but short-lived star reached the Earth, and he knows that in order to place this star as a beacon in the sky above Bethlehem, 2,000 years ago, his God had incinerated another world, had destroyed this ancient civilization "completely in the full flower of its achievement, leaving no survivors."[1]

Christians do not need to feel their faith challenged by a story like "The Star." After all, Clarke's enthralling story is just that, a piece of fiction. In addition, the Star of Bethlehem event appears only in the Gospel of Matthew. No contemporaneous historical accounts exist that suggest that any actual star shone in the skies above Bethlehem at the time of the birth of Jesus. Through the lens of modern science, the Star of Bethlehem can be understood as a real event, but one that involved a combination of astrological portents, rather than a bright light in the sky.[2] Viewed in this way, the wise men may have included astrologers who, following the Babylonian astrological tradition, predicted the coming of a great leader on the basis of the appearances of certain (not necessarily bright) planets in somewhat rare positions in the heavens relative to the positions of other planets. Alternatively, the story of the Star of Bethlehem might be one made up decades after the birth and death of Jesus. Matthew's story might have been devised in order to encourage potential believers, ones who accepted or practiced Babylonian astrology, to accept early claims for the divinity of Jesus. Some Christians, however, do accept the story of the Star of Bethlehem in Matthew's gospel as a written record of a historical event involving a bright star in the east that led the wise men to a stable in Bethlehem. For these Christians, an event like the one crafted by Clarke, were such a discovery to be made at some future time, would be a serious challenge to their beliefs, as this story illustrates in compelling fashion the potential conflict that can emerge when a scriptural story is interpreted as an actual, historical event and when that particular, historically-understood event is placed into the context of the possible existence of spiritual, intelligent extraterrestrials.

The understanding of most evangelical and fundamentalist Christians is that scriptural stories describe actual people and historical events that occurred at particular moments in time and at precise locations on a single planet, the one on which we live. As viewed through this literalist lens, most evangelical and fundamentalist Christian leaders argue quite forcefully that the Bible makes clear that extraterrestrial life does not exist. From this perspective, the only living, God-worshiping beings in the entire universe are humans, created by God, who live on Earth.

SOUTHERN BAPTISTS

"Evangelical Christians are committed to a supernatural worldview, which starts with the purposeful creation of the universe by God," writes Dr. R. Albert Mohler, Jr., in his 2005 essay "The Origin of Life: An Evangelical Baptist's View." A theologian and minister who since 1993 has served as the president of the Southern Baptist Theological Seminary in

---

[1] Clarke, A. C. (1974). The star. In *The nine billion names of God*. New York: New American Library.

[2] Molnar, M. R. (1999). *The star of Bethlehem: The legacy of the magi*. New Brunswick: Rutgers University Press.

Louisville, Kentucky, Mohler continues by saying, "Human beings are a special creation of God, made in His own image, and are granted important privileges, responsibilities and gifts that are to be used to God's glory." Mohler asserts that "most people recognize that there is something special—something unique—about human life. Human beings are set apart from other creatures and are the only self-conscious creatures."[3]

While we might question how Mohler knows that "most people" agree with this claim of human specialness, he very clearly means that humans are unique in the universe, not simply unique on the Earth. "If human beings are not made in the image of God," he continues, "and if the entire cosmos is nothing more than a freakish accident, morality is nothing but a mirage, and human beings—cosmic accidents that we are—are free to negotiate whatever moral arrangement seems best to us at any given time. Human life has no inherent dignity, morality has no objective basis, and we are alone in the universe to eat, drink and be merry before our bones join the fossil record and we pass from existence. ... The Christian doctrine of creation sets the stage for a comprehensive Christian view of life and human dignity. Without the doctrine of creation, Christianity is only one more artifact of an evolutionary process. The Christian affirmation represents the most significant intellectual challenge to evolutionary naturalism." In other words, evolutionary naturalism, which allows for the possibility that life could develop naturally anywhere in the universe, is the antithesis to the Christian doctrine of creation, in which all of God's attention is focused on a single type of being living on a single planet. Mohler agrees with Thomas Paine: one can believe in extraterrestrial life or one can believe in Christianity, but "the two beliefs cannot be held together in the same mind."[4] Mohler chooses Christianity, and in doing so would seem to be speaking, doctrinally at least, for most of the Southern Baptist faithful.

The Southern Baptist Convention, organized in 1845 in Augusta, Georgia, counts more than 16 million individual members in more than 45,000 churches, all of whom are expected to adhere to the religious convictions expressed in the most recent version of *The Baptist Faith and Message*, adopted in the year 2000. The year-2000 version of *The Baptist Faith and Message* replaces *The Baptist Faith and Message* adopted in 1963 while "respect[ing] the important contributions of the 1925 and 1963 editions of the *Baptist Faith and Message*."[5] The *Baptist Faith and Message* includes a number of statements that are important for understanding how Southern Baptists would approach the idea of extraterrestrial life.

According to the words in section *I. The Scriptures*, all Southern Baptists embrace the idea that the Holy Bible is "without any mixture of error" and that "all Scripture is totally true and trustworthy." Section *III. Man* begins with the words "Man is the special creation of God ... the crowning work of His creation." As part of the salvation beliefs of Southern Baptists (in section *IV. Salvation*), the *Baptist Faith and Message* explains, "There is no salvation apart from personal faith in Jesus Christ as Lord." Finally, in section *IX. The*

---

[3] Mohler, R. A. *The origin of Life: An evangelical baptist's view*. Retrieved August 5, 2005, from, http://www.npr.org/templates/story/story.php?storyId=4760816

[4] Paine, T. (1880). *The age of reason* (p. 38). London: Freethought Publishing Company.

[5] http://www.sbc.net/bfm/bfmcomparison.asp

Kingdom, the *Baptist Faith and Message* asserts, "The Kingdom of God includes ... His general sovereignty over the universe."[6]

The view of humanity as the "crowning work of His creation" and the belief that "there is no salvation apart from personal faith in Jesus Christ as Lord" is in opposition to the view, espoused by many in the last two centuries, for universalism. Timothy K. Beougher, the Billy Graham Professor of Evangelism and Church Growth at the Southern Baptist Theological Seminary, explains that in the universalist view, God will save all of humanity, even non-believers who might first enter hell, as even they will "sooner or later come out, having been brought to their senses and seeing their error in not acknowledging Christ."[7] The universalist view is clearly counter to the *Baptist Faith and Message* and suggests problems for extraterrestrial non-believers. Either extraterrestrial Others profess their "personal faith in Jesus Christ as Lord" or their souls perish in hell, assuming they have souls.

For clarification, the Chairman of the Committee on the *Baptist Faith and Message*, Adrian Rogers, wrote in 2000, "Given the pervasive influence of a postmodern culture, we are called to proclaim Jesus Christ as the only Savior and salvation in his name alone. Baptists thus reject inclusivism and pluralism in salvation, for these compromise the Gospel itself. Salvation comes only to those who call upon the name of the Lord, and come to personal faith in Jesus Christ as Savior."[8]

The simplest and easiest way to interpret the *Baptist Faith and Message* is that extraterrestrials do not exist; however, a slightly less human-centric interpretation is possible. Interpreted slightly less conservatively, the *Baptist Faith and Message* could be understood to mean that any and all intelligent beings, whether terrestrial or extraterrestrial, must be or must become Christians in order to earn or be given the gift of salvation. Based on this *Message* of exclusivity, one would have to assume, therefore, that extraterrestrials would be excluded, along with Jews, Muslims, Hindus, Buddhists and other Earth-bound non-Christians, unless the missionary work performed by those who already have a "personal faith in Jesus Christ as Lord" successfully converts the extraterrestrials to the Christian faith. Unless extraterrestrials are automatically consigned to damnation simply because they are extraterrestrials, their only path to salvation, according to the tenets of the Southern Baptist faith, is in the hands of Christian missionaries who have a duty to find them and bring Christianity to them. Since interstellar proselytizing will be difficult unless Southern Baptist missionaries find ways to overcome the laws of physics as currently understood, the assumption that extraterrestrials do not exist or that they are not eligible for salvation might be easier for a member of the Southern Baptist community to accept. Such an assumption also preserves the view that the humankind is the crowning achievement in God's creation.

For much of the twentieth century, Southern Baptist evangelical preacher Billy Graham (b. 1918) spoke to and for vast numbers of Southern Baptists and other conservative Christians in the United States. In 1976, despite the absence of scriptural evidence,

---

[6] http://www.sbc.net/bfm/bfm2000.asp

[7] Beougher, T. K. (1998, Summer). Are all doomed to be saved? The rise of modern universalism. *Southern Baptist Journal of Theology*.

[8] Rogers, A. *From the Chairman of the Committee on the Baptist Faith and Message.* http://www.sbc.net/bfm/bfmchairman.asp

Graham offered a firm answer about the existence of extraterrestrials: "I firmly believe there are intelligent beings like us far away in space who worship God. But we would have nothing to fear from these people. Like us, they are God's creation."[9]

Graham's understanding appears to be a bit different from that of the leading Southern Baptist theologians today. In Graham's understanding, his God is the creator of all life in the universe, and he presumes that 'people' beyond the Earth do exist. Graham's easy acceptance of intelligent life beyond the Earth can, however, be interpreted in a way that is consistent with the *Baptist Faith and Message* and also with the Baptist position against universalism: just as all humans on Earth are God's creation but only some humans are Christians, all intelligent beings in all places are God's creation, but only some of them are, can be or will become Christians.

Graham does not concern himself with an explanation as to how extraterrestrials will learn about or why they need to know about Christ and Christianity, or even if becoming Christian is possible for them, but his and the current Southern Baptist views are clear: humans on Earth are privileged to know about Christ and His message of salvation. Extraterrestrial beings, if they exist, can only hold out hope that somehow, someday Christians on Earth will transmit their Christian knowledge across space and time so that the extraterrestrials can be saved. In this sense, the discovery of rational, intelligent extraterrestrial beings would not threaten the faith of Southern Baptists; rather, those beings would offer another, albeit a logistically formidable, opportunity for missionary work. In fact, one could surmise from Southern Baptist beliefs that humans should be aggressively searching for signs of extraterrestrial life, because unless earthly Christians are able to reach out and convert them to Christianity, any such extraterrestrial beings are doomed to everlasting punishment in hell.

CREATIONISTS

As explained by Ted Peters in his book *Science, Theology, and Ethics*,[10] fundamentalist Christians have at least three very good reasons to disagree with Billy Graham's position that "there are intelligent beings like us far away in space." The first of these is that unless one chooses to interpret the "sheep" in a passage such as John 10:16 ("I have other sheep that do not belong to this fold") as extraterrestrials, the Bible does not reveal the existence of extraterrestrials; therefore, extraterrestrials do not exist. This argument is consistent with biblical literalism, which is a foundation of modern fundamentalist thinking. Of course, this view fails an obvious test of logic, as many things are never explicitly mentioned in scripture. Some obvious early twenty-first century examples of objects unmentioned in scripture include computers, cell phones, airplanes, dishwashers, watches, milkshakes, vaccines and antibiotics. The number of things one could identify that are not mentioned in the Bible is limited only by one's imagination, time and experiences, but

---

[9] From *National Enquirer* (1976, November 30), as quoted in Peters, T. (2003). *Science, theology, and ethics* (p. 126). Ashgate science and religion series.

[10] Peters, T. (2003). *Science, theology, and ethics* (pp. 130–131). Ashgate science and religion series.

even fundamentalists would struggle to argue that all of these biblically-unmentioned modern devices do not exist.

The second and third of Peters's reasons as to why fundamentalists would disagree with Billy Graham's belief in the existence of extraterrestrial beings are both slippery slope arguments. If one accepts the existence of extraterrestrials, then one will find oneself sliding rapidly down one slippery slope toward accepting evolution and down a second slippery slope toward embracing the possibility of a non-Christian path for earning or being granted salvation.

Why? Extraterrestrials, as imagined in our popular culture, are technologically more advanced than humans. Whether they are coming to destroy, enslave, enlighten or merely visit us, they have more control over the physical world than we have. Either God chose to create humans as stupid creatures, inherently disadvantaged both intellectually and technologically in comparison to our extraterrestrial neighbors, or these other species of sentient life, given more time or different living conditions, have evolved past our level of control over our environment. Unless we embrace the concept of a God who intentionally created us as lesser beings in comparison to extraterrestrials, then we are trapped into accepting the concept of evolution, at least on other worlds. If evolution is possible on other worlds, for other life forms, then we too might have the potential to evolve. Therefore we may have evolved, over a significant period of time, from less advanced life forms. Since opposition to evolution is a *sine qua non* of modern fundamentalist Christianity, opposing the very existence of extraterrestrial life is necessary for a logically consistent, fundamentalist belief system.

The second slippery slope leads toward accepting a path to salvation or even eternal life via extraterrestrial wisdom rather than through Christian salvation. Maybe extraterrestrials have conquered death through their advanced technologies. If so, Christianity would no longer be the only path toward eternal life. In the words of Peters, this would be 'demonic.'

Creationist writer Jonathan D. Sarfati firmly embraces the position that we are alone in the universe. In his essay "Bible Leaves No Room for Extraterrestrial Life," he writes, "Scripture strongly implies that no intelligent life exists elsewhere. ... The Bible says nothing to indicate that God created life anywhere but Earth."[11] Sarfati, the co-editor of *Creation* magazine and a prolific creationist writer who defends a literal biblical timeline in his writings, defends the uniqueness of life on Earth. "Earth was created to be home for creatures made in God the Creator's image. It was on Earth that the first human couple rebelled against its creator and brought the cosmos under His curse. Thus it would have affected Martians, Vulcans, Klingons and any other being in the universe. The second person of the holy trinity incarnated on Earth alone, took on human nature, died for the sins of those with whom He has the kinsman-redeemer relationship, then ascended to the right hand of God the Father. He did not take on Vulcan or Klingon nature, and He will have only one bride—the church—for all eternity. It would therefore seem hard to reconcile intelligent life on other worlds with the doctrine of the incarnation.

---

[11] Sarfati, J. D. (2004, March). Bible leaves no room for extraterrestrial life. *Science and Theology News*.

It would also seem odd for God to create microscopic life on other planets, but we should not be dogmatic on this."

After offering these strictly biblically-based arguments, Sarfati offers arguments premised on his claimed expertise in biology, chemistry and physics paired with his personal confidence in his ability to correctly interpret the Bible: "… despite spending millions of dollars, NASA and others have not found the slightest proof for life anywhere but Earth. Behind the search is the metaphysical assumption that life evolved from nonliving chemicals on Earth, so there is no reason it couldn't evolve elsewhere. Chemical evolution has such major hurdles that if life were found on Mars, the most reasonable assumption is that it came from Earth somehow. Scripture mentions nothing about biology outside Earth, but looking at the big picture of the Bible, it seems hard to reconcile it with extraterrestrial intelligence."

Of course, the absence of evidence of any natural phenomenon is not evidence of the absolute, universal absence of that phenomenon, even if Sarfati prefers to interpret "the absence of the slightest proof for life anywhere but Earth" after only three decades of limited scientific investigations as evidence in support of his arguments against evolution and for the uniqueness of life on Earth.

Sarfati's views are echoed in the words of Ken Ham. Ham is the President of "Answers in Genesis" in the United States, which self-identifies as "the world's largest apologetics (i.e., Christianity-defending) ministry, dedicated to enabling Christians to defend their faith, and to proclaim the gospel of Jesus Christ effectively."[12] Ham's *Answers* magazine regularly wins awards from the Evangelical Press Association; he regularly writes in his blog "Around the World with Ken Ham." On June 18, 2008, Ham blogged, "I like to remind people of Psalm 115: 16, 'The heaven, even the heavens, are the LORD'S; But the earth He has given to the children of men.' Scripture certainly makes earth the center stage. The heavens are there to 'declare the glory of God,' but the earth was made for humans to inhabit. So it seems even from these passages, one would not expect life in outer space— only the earth was made specially for intelligent physical beings to dwell on. Also, the Bible makes it clear that the *whole* of creation groans because of sin (Romans 8:22)—and that Jesus stepped into history on earth to become a human (the God man)—a perfect man, but God, so He could die on a cross, be raised from the dead and offer a free gift of salvation. Jesus remains the 'God man,' as he is our Savior. Jesus did not become a 'Martian' or a 'Klingon' or some other being—he became a human (as God)."

Ham continues by asserting, unequivocally, the uniqueness of intelligent life on Earth. "So, it wouldn't make any sense for there to be intelligent beings like us on other planets— they would be suffering from the effects of sin but can't have salvation, as only descendants of Adam can be saved. One day the whole universe will be wound up—the judgment by fire—and there will be a New Heavens and Earth. I always say that there can't be intelligent life like us on other planets—the Bible does not say there is or is not animal or plant life on other planets—but I highly suspect not."[13]

---

[12] http://blogs.answersingenesis.org/blogs/ken-ham/about/. Accessed 10 July, 2012.
[13] Ham, K. Retrieved from http://blogs.answersingenesis.org/blogs/ken-ham/2008/06/18/speaking-at-the-pentagon/. Accessed July 12, 2012.

Walt Brown, the Director of the Center for Scientific Creation and author of *In the Beginning: Compelling Evidence for Creation and the Flood*, agrees with Sarfati and Ham. Brown argues that those who believe life exists on distant planets are using flawed reasoning. He sums up the argument in favor of a universe teeming with life in this way: "Live evolved on Earth. Because the universe is so immense and contains so many heavenly bodies, life independently evolved on other planets as well." He dismisses this logic because the argument "assumes that life evolved on Earth," yet "overwhelming evidence shows that life is so complex it could not have evolved—anywhere!" Brown explains, further, that the hostile environments that exist everywhere in the universe except on Earth make life impossible: "Conditions outside Earth are more destructive than almost anyone suspected before space exploration began: deadly radiation, poisonous gases, extreme gravitational forces, gigantic explosions, and the absence of the proper atmospheres and chemical elements. Just the temperature extremes in outer space would make almost any form of life either so hot it would vaporize or so cold it would be completely rigid, brittle, and dead."[14]

The scientific evidence against the existence of life beyond the Earth is, according to Brown, quite strong: "If life evolved in outer space as easily as some people believe, many extraterrestrial "civilizations" should exist, especially on planets around stars that evolutionists claim are older than our Sun. Some civilizations should even be technologically superior to ours, would have recognized that earth has abundant life, and would have tried to reach us. Any superior civilization within our galaxy would probably have already explored our solar system, at least with robots. Because we have no verifiable evidence of any of this, intelligent extraterrestrial life probably does not exist, certainly within our Milky Way Galaxy."[15] As does Sarfati, Brown invokes the popular 'absence of evidence is evidence of absence' argument as proof-positive for the uniqueness of humanity.

"Could God have created life elsewhere?" Brown asks. "Certainly," is his answer, "but the Bible is largely silent on this subject." He then offers three Bible verses that he says "suggest that conscious, rational life is unique to Earth."[16] First, "Romans 8:22 states 'the whole creation groans and suffers' because of Adam's sin. This would be a strange statement," Brown says, "if humanlike beings existed in outer space, because it would mean that although not descended from Adam, they suffer because of his sin." Second, "Romans 5:12 tells us, 'through one man [Adam] sin entered the world.' The Greek word we translate as 'world' is kosmos, which generally means the entire universe. Again, if intelligent beings exist beyond Earth," says Brown, "they would be suffering for Adam's sin." And third, "Genesis 1:14 states that the heavenly bodies were made 'for signs, and for seasons, and for days and years.' It does not say that they were created as habitats for other creatures."[17]

---

[14] Brown, W. (2008). *In the beginning: Compelling evidence for creation and the flood* (8th ed., online edition). Center for Scientific Creation. "Is There Life In Outer Space?" Retrieved from http://www.creationscience.com/onlinebook/

[15] Ibid.

[16] Ibid.

[17] Ibid.

In his book *Taking Back Astronomy*,[18] Jason Lisle, the planetarium director at the Creation Museum in Kentucky, repeats the arguments that "the idea of extraterrestrial life stems largely from a belief in evolutionism" and that "the notion of alien life does not square well with Scripture." Lisle asks, rhetorically, "Where does the Bible discuss the creation of life on the 'lights in the expanse of the heavens?' There is no such description, because the lights in the expanse were not designed to accommodate life." Lisle also repeats the argument that "Intelligent alien beings cannot be redeemed! God's plan of redemption is for human beings: those descended from Adam." His reasoning for this position is clear: "they are not blood relatives of Jesus, and so Christ's shed blood cannot pay for their sins. One might at first suppose that Christ also visited their world, and lived and died there as well, but this is anti-biblical. Christ died once for all (1 Peter 3:18; Hebrews 9:27, 10:10). Jesus is now and forever both God and man; but He is not an 'alien.' One might suppose that alien beings have never sinned, in which case they would not need to be redeemed, but then another problem emerges: they suffer the effects of sin, despite having never sinned. Adam's sin has affected all of creation—not just mankind. Romans 8:20–22 makes it clear that the entirety of creation suffers under the bondage of corruption. These kinds of issues highlight the problem of attempting to incorporate an antibiblical notion into the Christian worldview."

Henry Morris, writing on the Institution for Creation Research website,[19] dismisses as "blasphemously arrogant, as well as utterly foolish" any speculations about extraterrestrial life. Such activities are "nothing but expressions of man's rebellion against his Creator." He continues, "There is a lot of extraterrestrial life, of course. The Bible calls them angels! These are specially created beings 'sent forth to minister for them who shall be heirs of salvation' (Hebrews 1:14). But the astro-scientists will not find any angels with their telescopes or space probes. Neither will they find any other humans or humanoids out there. Physical life was specially created, and Earth was created uniquely to support that life. The stars were created for other purposes, not yet revealed. 'The heaven, even the heavens, are the LORD'S: but the earth hath He given to the children of men' (Psalm 115:16)."

Creation Ministries International CEO Gary Bates writes, "the biblical objection to ET is not merely an argument from silence. Motor cars, for example, are not a salvation issue, but we believe that [the existence of] sentient, intelligent, moral-decision-capable beings is, because it would undermine the authority of Scripture. In short, understanding the big picture of the Bible/gospel message allows us to conclude clearly that the reason the Bible doesn't mention extraterrestrials (ETs) is that there aren't any. Surely, if the earth were to be favoured with a visitation by real extraterrestrials from a galaxy far, far away, then one would reasonably expect that the Bible, and God in His sovereignty and foreknowledge, to mention such a momentous occasion, because it would clearly redefine man's place in the universe."[20]

---

[18] Lisle, J. (2006). *Taking back astronomy.* Master Books. Retrieved October 7, 2013, from http://www.answersingenesis.org/store/product/taking-back-astronomy/

[19] Morris, H. *The heavens are the lord's.* Retrieved from http://www.icr.org/article/20964/ and *The Stardust Trail.* Retrieved July, 11, 2012, from http://www.icr.org/article/stardust-trail/

[20] Bates, G. *Did God create life on other planets?* Retrieved from http://creation.com/did-god-create-life-on-other-planets

For these fundamentalist leaders, all of whom have very large numbers of followers, the discovery of extraterrestrial life would be devastating. Theologically, they are totally unprepared for and unwilling to entertain a universe that includes any advanced life forms other than humans on planet Earth. Humanity is, they believe, God's singular focus. The theological foundation for their Christianity rests on the idea that the incarnation and the resurrection of Jesus were given by God exclusively to and for humans. Christianity, as understood and practiced by the members of these religious groups, is the universal religion, and extraterrestrial beings cannot and do not exist.

13

From Christian Roots

> *Sin has never entered here.*
>
> attributed to Ellen G. White in a letter of Mrs. Truesdail
> (Mrs. Truesdail's letter of Jan. 27, 1891, as quoted
> in Loughborough, J. N. (1905). *Great second
> advent movement* (p. 213) (reprinted by
> Adventist Pioneer Library, 1992)).

The religions of Unitarian Universalism, the Religious Society of Friends (Quakers), Christian Scientism, Seventh-day Adventism and Jehovah's Witnesses all have roots in protestant Christianity. All, however, have grown in very different directions from those original roots. The belief systems of these five religions cover a broad spectrum in terms of how different religions might offer a variety of approaches for handling the discovery of extraterrestrial life. Though neither the Unitarian Universalists nor the Quakers appear to have anything to say, explicitly, about life in the universe beyond the Earth, they both fully accept the discoveries of modern science. Thus, followers of both religions likely would easily embrace the discovery of life beyond Earth. In contrast, Christian Scientism rejects significant aspects of modern science; followers of Christian Science even nominally reject the notion that the human body is real. Given the Christian Science belief that the human mind is real but the body is not, Christian Science believers probably would reject claims that extraterrestrial beings are anything other than illusions. In stark contrast, the prophetess of Seventh-day Adventism, Ellen G. White, claimed to have encountered living beings on Saturn, and so extraterrestrial life is intimately associated with her prophesies and with Seventh-day Adventist beliefs. Finally, Jehovah's Witnesses, like many fundamentalist Christians, believe in the absolute centrality of Earth in God's purpose and for this reason most likely would reject the possibility that beings created by God could exist anywhere else.

D.A. Weintraub, *Religions and Extraterrestrial Life: How Will We Deal With It?*,
Springer Praxis Books, DOI 10.1007/978-3-319-05056-0_13,
© Springer International Publishing Switzerland 2014

UNITARIAN UNIVERSALISM

Both the Unitarians and the Universalists emerged out of the Christian tradition, but each with important doctrinal differences from more conservative or traditional Christian faith traditions. The original Unitarian congregations, which formed in Transylvania in the sixteenth century, emerged from a faith group of Christians who rejected the Christian doctrine of the Trinity and advocated following but not worshiping Jesus. The Universalists began as a faith group of Christians, officially organized in the eastern United States and Canada in 1793, who believed that all people of all denominations will earn or be offered salvation.

In 1961, the Universalist Church of America and the American Unitarian Association consolidated to form the Unitarian Universalist Association. The modern, unified Unitarian Universalist Church, which includes about 800,000 members, worldwide, is pluralistic, embracing "theist and atheist, agnostic and humanist, pagan, Christian, Jew, and Buddhist."[1] Unitarian Universalists have no single scripture and no single set of beliefs that all followers accept. They do draw from Jewish, Christian, eastern, humanist and spiritual teachings and "heed the guidance of reason and the results of science" while promoting "a free and responsible search for truth and meaning."[2] Clearly, such principles would be compatible with the discovery of extraterrestrial life or the data-driven conclusion that, in at least a substantial part of our galaxy, we are alone.

RELIGIOUS SOCIETY OF FRIENDS (QUAKERS)

The Religious Society of Friends, commonly known as the Quakers, was founded by George Fox (1624–1691 C. E.) in England in the middle of the seventeenth century, and now includes about 400,000 followers, worldwide. At the time of its founding, The Society of Friends offered an alternative approach to Christianity within the emerging Protestant tradition. Today, some Quakers hold traditional Christian beliefs while others come from a variety of religious backgrounds and do not consider themselves Christians. They "share a way of life, not a set of beliefs,"[3] and have in common a commitment to the idea that "every person is known by God and can know God in a direct relationship,"[4] without the mediating help of ordained leaders. Quakers espouse a commitment to principles of truth, equality and peace and have a long tradition of engagement with the world, including acceptance of the scientific understanding of the age.

With such broad acceptance of science and of beliefs from all religions, Quakers are unlikely to have any preconceived notions as to whether extraterrestrials exist. The existence of rational beings beyond the Earth would have no bearing on an individual Quaker's

---

[1] Harris, M. W. *Unitarian universalist origins: Our historic faith.* Retrieved from http://www.uua.org/publications/pamphlets/introductions/151249.shtml

[2] *Our unitarian universalist principles.* Retrieved from http://www.uua.org/beliefs/principles/index.shtml

[3] *Introducing Quakers.* Retrieved from http://www.quaker.org.uk/intro-quakers

[4] *The Quaker way.* Retrieved from http://www.fgcquaker.org/explore/quaker-way

personal relationship with God, and if extraterrestrials do exist, Quakers would not presume to impose their religious ideas on them, let alone assert that their personal religious beliefs and concepts make sense for beings from other worlds.

SEVENTH-DAY ADVENTISTS

The Seventh-day Adventist Church, which claims a worldwide membership of more than 17 million faithful, traces its roots to the teachings of Adventist preacher William Miller (1782–1849 C. E.), who predicted that the Second Coming, or advent, of Christ would occur sometime between March 21, 1843 and March 21, 1844. When Christ failed to appear within this temporal window, Miller issued a new prediction for the advent of Christ, October 22, 1844, which also ended in disappointment for him and his followers. Beginning in 1844, one youthful Adventist, Ellen Harmon (later Ellen G. White) claimed to have visions that led her and her future husband James White (they married in 1846) to co-found the Seventh-day Adventist Church in 1860. After the founding, the new church proclaimed her as a prophet. Thereafter, Ellen G. White continued in a leadership role with the Seventh-day Adventist Church for several decades.

Seventh-day Adventists hold to a set of fundamental beliefs that include the idea that the Holy Scriptures, these being the Old and New Christian Testaments, are the "Written Word of God" and that God's acts of creation are related accurately in these writings. As understood by Seventh-day Adventists, these beliefs lead directly to the knowledge that God created the heaven and the earth in 6 days and that the second coming of Christ is imminent. Believers are "exhorted to be ready at all times."[5] Seventh-day Adventists believe that Christ reigns with His saints in heaven and that true believers will be resurrected and will receive sanctuary in heaven after the second coming of Christ. While the language of Seventh-day Adventist Church writings are unclear as to whether heaven is an actual location in the physical universe, Ellen G. White's writings suggest that it likely is. She even proclaims that the location of the door to heaven is in the great nebula in Orion, for she saw that door in a vision she had on December 16, 1848. In her words, "we could look up through the open space in Orion, whence came the voice of God. The Holy City will come down through that open space."[6]

In 1847, at a time when James White was struggling to determine for himself whether his wife's visions were divinely inspired or "from the devil," he wrote a pamphlet *A Word to the "Little Flock"*. In this pamphlet, he includes the following description of one of her visions: "At our conference in Topsham, Maine, last Nov., Ellen had a vision of the handiworks of God. She was guided to the planets Jupiter, Saturn, and I think one more. After she came out of vision, she could give a clear description of their moons, etc."[7]

Because, he writes, "she knew nothing of astronomy … before she had this vision," this vision convinced James White that Ellen G. White's visions were divinely inspired. Another Adventist, Mrs. Truesdail, "who was present on the occasion of the giving of

---

[5] http://www.adventist.org/beliefs/fundamental/index.html

[6] White, E. G. (1882). Shaking the powers of heaven. In *Early writings* (p. 41). Review and Herald Publishing Association. Retrieved from https://egwwritings.org/

[7] White, J. S. (1847). *A word to the "Little Flock"*. Brunswick, ME.

[this] vision,"[8] also wrote about her remembrances of this experience. "The Spirit of God rested upon us," wrote Mrs. Truesdail. "We soon noticed that she [Ellen G. White] was insensible to earthly things. This was her first view of the planetary world. After counting aloud the moons of Jupiter, and soon after those of Saturn, she gave a beautiful description of the rings of the latter. She then said, 'The inhabitants are a tall, majestic people, so unlike the inhabitants of earth. Sin has never entered here.'"[9] Such ideas about inhabitants of other planetary worlds in our solar system were very much in keeping with ideas that were prevalent in both the astronomy and lay communities in the early and mid-nineteenth century, and the idea that of all created, sentient beings in the universe only we humans on Earth are sinful also matches popular thought at the time. Except for James White's assertion that "she knew nothing of astronomy ... before she had this vision," we have no way of knowing whether Ellen G. White was aware of any of those astronomical ideas prior to the time of her reported vision. Seventh-day Adventists, of course, accept James White's word on the matter and embrace Ellen G. White's vision, in which she saw the tall, pure inhabitants of Saturn, as divinely inspired.

Ellen G. White consistently mentioned her views on extraterrestrials and about the absence of sin in worlds beyond the Earth. In her *Early Writings*, published in 1882, in a section entitled 'God's Love for His People,' she writes, "The Lord has given me a view of other worlds.... The inhabitants of the place were of all sizes; they were noble, majestic, and lovely. ... I asked one of them why they were so much more lovely than those on the earth. The reply was, 'We have lived in strict obedience to the commandments of God, and have not fallen by disobedience, like those on the earth.'"[10]

A decade later, in her 1890 book *The Story of Patriarchs and Prophets*,[11] White describes how sin entered the world through Lucifer's conflict with God. In these writings, White repeatedly refers to inhabitants of other worlds. First, she writes that Lucifer began to have "doubts concerning laws that governed heavenly beings, intimating that though laws might be necessary for the inhabitants of the worlds, angels ... needed no such restraint." Then we learn how God acted in order to counter the claims of Lucifer: "It was therefore necessary to demonstrate before the inhabitants of heaven, and of all the worlds, that God's government is just, His law perfect."

Beliefs in extraterrestrial life and in the uniquely sinful state of humanity are not clearly identified or advertised by the modern Seventh-day Adventist Church as part of the fundamental beliefs of the Adventist faithful. An official statement of the Seventh-day Adventist Church, however, issued by its General Conference on June 30, 1995, entitled *A Statement of Confidence in the Spirit of Prophecy*,[12] establishes the Church's current position with regard to the life and ministry of Ellen G. White. Ellen G. White, according

---

[8] Loughborough, J. N. (1905). *Great second advent movement* (p. 212) (reprinted by Adventist Pioneer Library, 1992).

[9] Mrs. Truesdail's letter of Jan. 27, 1891, as quoted in Loughborough, J. N. (1905). *Great second advent movement* (p. 213) (reprinted by Adventist Pioneer Library, 1992).

[10] White, E. G. (1882). God's love for his people. In *Early writings* (p. 39). Review and Herald Publishing Association. Retrieved from https://egwwritings.org/

[11] White, E. G. *The story of patriarchs and prophets* (p. 37, 42). Retrieved from http://www.whiteestate.org/books/pp/pp1.html

[12] http://adventist.org/beliefs/statements/main-stat24.html

to *A Statement of Confidence*, fulfilled God's promise to provide "the remnant of the church with the 'spirit of prophecy.'" The *Statement of Confidence* continues, "we believe she did the work of a prophet. ... As Seventh-day Adventists, we believe ... that her writings carry divine authority." This firm support for accepting Ellen G. White's role as a prophet is reinforced through the 28 Fundamental Beliefs held by Seventh-day Adventists, the eighteenth of which is The Gift of Prophecy. The Church's description of The Gift of Prophecy states, "This gift is an identifying mark of the remnant church and was manifested in the ministry of Ellen G. White. As the Lord's messenger, her writings are a continuing and authoritative source of truth which provide for the church comfort, guidance, instruction, and correction."[13]

Though issues concerning the existence and lack of sinfulness of extraterrestrial beings may not be parts of the everyday, twenty-first century concerns of practicing Seventh-day Adventists, Ellen G. White's views on extraterrestrial life continue to be formally embraced as part of the foundational doctrines of the Seventh-day Adventist movement. Thus, even though modern Seventh-day Adventist literature carefully avoids any mention of Ellen G. White's views on extraterrestrial life, these ideas, which are clearly put forward in her visions and writings, most certainly have not been rejected. For Seventh-day Adventists, the discovery by astronomers of life on other planets would not be shocking and would perhaps even bolster their faith in all of Ellen G. White's writings, including, perhaps most importantly, that humans are the only sinful beings in the universe.

JEHOVAH'S WITNESSES

Jehovah's Witnesses also emerged from the nineteenth-century Adventist tradition. With regard to extraterrestrial life, however, Jehovah's Witnesses land at the opposite end of the spectrum from Seventh-day Adventists. The literature and beliefs of Jehovah's Witnesses address the concept of extraterrestrials only indirectly. In doing so, Jehovah's Witnesses assert the absolute centrality of humanity on Earth to God's plan for the universe—this idea is consistent with Seventh-day Adventist beliefs—and consequently the impossibility that life beyond the Earth can exist.

The original leaders of this movement called themselves the Zion's Watch Tower Tract Society. Beginning in 1879, under the leadership of Charles Taze Russell, they published the magazine *Zion's Watch Tower and Herald of Christ's Presence*. The movement renamed itself Jehovah's Witnesses in 1931 and now claims nearly eight million adherents worldwide.

Jehovah's Witnesses believe the Bible, specifically the *New World Translation of the Holy Scriptures*, a 1961 (and subsequently revised) translation of the New and Old Testaments of the Christian tradition commissioned by the New World Bible Translation Committee and published by the Watch Tower Bible & Tract Society, is inspired by God and is "incomparable in its reliable prophecy, historical and scientific accuracy, internal

---

[13] http://www.adventist.org/fileadmin/adventist.org/files/articles/official-statements/28Beliefs-English.pdf

harmony, and practical guidance."[14] According to the beliefs of Jehovah's Witnesses, the words in the Bible are not merely inspired by God; rather, they were dictated by God: "God by means of his holy spirit guided the Bible writers to write only what He wanted them to write."[15]

Charles Taze Russell preached that Christ had returned to Earth in 1874, but in doing so Christ returned both invisible and undetected. Russell predicted that all believers would be called to heaven in 1914. They were not. In 1966, half a century later, Jehovah's Witnesses leaders announced that the end of the world would occur in 1975. The world did not end. Jehovah's Witnesses now only say that they expect that the end of days is imminent; they do not regard Russell's writings, in this regard, as scriptural.

Jehovah's Witnesses do believe that God, who is called Jehovah in this religious tradition, created everything in existence. Jesus Christ, they believe, is the Son of God, superior to the angels but a lesser being than God. By denying the traditional Christian idea of the Trinity, Jehovah's Witnesses depart radically from the various Christian traditions; however, while for this reason many Roman Catholics and Protestants refuse to identify Jehovah's Witnesses as Christians, the Jehovah's Witnesses themselves assert that they "are proud to be called Christians."[16]

Jehovah's Witnesses believe that at the end of days "144,000 men and women resurrected to life in heaven will rule over the earth." Humanity on Earth will then enter a 1,000-year period known to Jehovah's Witnesses as Judgment Day, which "will affect those living on 'the *inhabited earth*.' Those receiving favorable judgment will live on earth and will enjoy everlasting life in perfect conditions." All others will "be permanently destroyed. … Judgment Day will thus be a part of the accomplishment of God's purpose to undo all the effects of the original rebellion against God in the garden of Eden."[17] When Judgment Day comes, "wicked human society" will end, but the physical planet Earth itself and "godly human society will last forever."[18]

From the standpoint of Jehovah's Witnesses theology, Earth is the sole focus of God's attention, as only those living on Earth might be saved. The heavens, that is the stars, the Sun and the Moon, according to this theology, were created for a very clear purpose, to help humans track time. This purpose is clearly spelled out in Genesis, as has been explained in the article "Do the Stars Affect Your Life?" published in the Jehovah's Witnesses journal *Awake!* "The Bible is not a scientific textbook, spelling out every detail about the human body and the universe. However, it does explain the purpose for which Jehovah created the heavenly bodies. Genesis 1:14, 15 states: 'God said, "I command lights to appear in the sky and to separate day from night and to show the time for seasons … I command them to shine on the earth.""[19]

---

[14] This means everlasting life. (2012, June 1). *The Watchtower*, p. 9.

[15] *The Bible really is God's inspired word*. Retrieved from http://www.jw.org/en/publications/magazines/wp20100301/bible-inspired-of-god/

[16] *Jehovah's witnessess—who are we?* Retrieved from http://www.jw.org/en/

[17] "What is Judgment Day?" (italics in original text). Retrieved from http://www.jw.org/en/publications/magazines/g201001/what-is-judgment-day/

[18] Will the earth come to an end? (2012, February 1). *The Watchtower*, p. 25.

[19] *Do the stars affect your life?* Retrieved from http://www.jw.org/en/publications/magazines/g201210/do-the-stars-affect-your-life/

For Jehovah's Witnesses, the Word of God in the Bible very clearly offers evidence that God created the entire universe in order to have a place, the Earth, that would be the eternal home for those who proved themselves faithful to Jehovah. The heavens do not provide abodes for extraterrestrial beings; the heavenly bodies exist as an enormous clock and calendar. While the Bible, in the form of the *New World Translation of the Holy Scriptures*, does not explicitly rule out intelligent beings on other planets, according to the beliefs of Jehovah's Witnesses such beings are not part of God's purpose. Since the God worshiped by Jehovah's Witnesses is purposeful, when reading between the lines the message becomes clear: we are the only created, intelligent beings in the universe. The discovery of advanced, rational creatures on other worlds most likely would greatly surprise and trouble the most fervent Jehovah's Witnesses.

THE CHURCH OF CHRIST, SCIENTIST

The Church of Christ, Scientist, known less formally as the Christian Science Church, was founded by Mary Baker Eddy in 1879. (The Church of Christ, Scientist is a different organization than the many autonomous, undenominational Church of Christ houses of worship and from the denominational United Church of Christ.) According to its own literature, Christian Science is "deeply Christian."[20] Christian Science, however, offers a view of reality that is more Platonic and eastern than is normally found in western religions. The written foundations for Christian Science are the Christian Bible and Eddy's textbook *Science and Health with Key to the Scriptures*, first published in 1875.

For followers of Christian Science, the human mind is real but the human body is an illusion. For this reason, according to Christian Science beliefs, human illness is illusory. Since the mind can discover through prayer that it is healthy despite the physical ailments that appear to afflict the body, "many Christian Scientists find that prayer consistently results in healing."[21]

Mary Baker Eddy and Christian Science have nothing to say directly about extraterrestrial life. Eddy did, however, pen these words in the chapter 'Genesis' in *Science and Health*: "Divine Science deals its chief blow at the supposed material foundations of life and intelligence. It dooms idolatry. A belief in other gods, other creators, and other creations must go down before Christian Science."[22] Eddy's words can be understood to suggest that a follower in Christian Science would believe that no other intelligent beings exist. In addition, as a natural extension of the Christian Science belief that the human body is merely an illusion, we can surmise that any extraterrestrials we might discover or meet could also be dismissed, like illnesses, as illusions. Whether extraterrestrials do or do not exist, followers of the Church of Christ, Scientist appear to have no theological issues that will keep them up at night worrying.

---

[20] From *Common confusions*. christianscience.com/what-is-christian-science

[21] From *How do prayer and healing work*. christianscience.com/what-is-christian-science

[22] Eddy, M. B. *Science and health with key to the scriptures* (Chap. 15). Retrieved from http://christianscience.com/read-online/science-and-health

14

Mormonism

> We are not the only people that the Lord has created.
> We have brothers and sisters on other earths.
>
> Joseph Fielding Smith

From its beginnings and continuing through today, Mormon theology and belief have had an intimate connection with ideas about extraterrestrial life. More precisely, the existence of humanlike life on worlds beyond the Earth is central to Mormon theology and belief. According to Mormon belief, human bodies are mortal. Human spirits, however, are eternal and have been, for all of eternity until the present, lesser intelligences than God himself. God, however, has created other worlds, and humans who become exalted, who earn eternal life, may be living as gods in human form on those worlds. These ideas can be found in a combination of scriptural texts, in the authoritative teachings of Mormon prophet Joseph Smith, Jr., and in the words and writings of Mormon leaders, including Brigham Young in the nineteenth century and Joseph Fielding Smith and Neal A. Maxwell in the late twentieth century.

Mormonism originated in the 1820s in western New York after Joseph Smith, Jr. (1805–1844 C. E.) had a series of visions in which he believed that he was instructed to re-establish the true Christian church. Formally known as The Church of Jesus Christ of Latter-day Saints and headquartered in Salt Lake City, the largest branch of Mormonism now counts 15 million followers, worldwide. Mormonism offers a religious foundation that strongly supports the idea that intelligent life, in particular humanlike life, exists beyond the Earth. Some of these ideas are found in the foundational scriptural texts of Mormonism. In addition, because "a central principle of Mormonism is belief in ongoing revelation,"[1] other

---

[1] Bushman, R. L. (2008). *Mormonism: A very short introduction* (p. 3). Cary, NC: Oxford University Press.

D.A. Weintraub, *Religions and Extraterrestrial Life: How Will We Deal With It?*,
Springer Praxis Books, DOI 10.1007/978-3-319-05056-0_14,
© Springer International Publishing Switzerland 2014

important concepts are found in the writings and teachings of past and current Mormon leaders, whose ideas, according to Mormon beliefs, are divinely inspired.

Mormons hold a broad set of written documents as sacred scripture. These include the *Hebrew Bible* or *Old Testament*, which Mormonism shares with Judaism, Catholicism and the many Protestant faiths, and the *New Testament*, which Mormonism shares with Catholicism and the Protestant faiths. Mormon scripture also includes the foundational text the *Book of Mormon*, which is considered to be "a volume of holy scripture comparable to the Bible,"[2] the *Doctrine and Covenants* ("a collection of divine revelations and inspired declarations," most of which were "received through Joseph Smith, Jr., the first prophet" of Mormonism[3]) and the *Pearl of Great Price* ("a selection of materials ... produced by the Prophet Joseph Smith," that includes the *Book of Moses* and the *Book of Abraham*, among other texts[4]).

The most important non-scriptural documents for understanding Mormon cosmological beliefs, which in turn provide insights into Mormon views on the possibility of extraterrestrial life, is Joseph Smith, Jr.'s *King Follett Sermon*. This sermon, which Smith delivered at a funeral in 1844 for his close friend King Follett, is described on the official website of the Church of Jesus Christ of Latter-day Saints as "one of the classics of Church literature."[5] While the King Follett Sermon is not part of the body of sacred scripture for Mormons, Smith told his followers when giving this discourse that he was "inspired by the Holy Spirit,"[6] and the summary of Mormon pre-human history and cosmology spelled out by Smith in this discourse is widely accepted in more recent, inspired writings of Mormon prophets.

The starting point for Mormon cosmology is in the Doctrines and Covenants, which teaches that "the elements are eternal" (Doctrines and Covenants 93:33) but God is not. In the King Follett Sermon, Smith tells his followers, "We have imagined and supposed that God was God from all eternity. I will refute that idea... He was once a man like us; yea, that God himself, the Father of us all, dwelt on an earth, the same as Jesus Christ Himself did."[7] According to Richard Bushman, an historian of Mormonism and one of three general editors of *The Joseph Smith Papers*, Joseph Smith, Jr.'s "greatest theological departure [from traditional Christianity] was to state that God was of the same order of being as humans."[8]

---

[2] 'Introduction' to the *Book of Mormon*. Retrieved from http://www.lds.org/scriptures/bofm/introduction

[3] 'Introduction' to the *Doctrine and covenants*. Retrieved from http://www.lds.org/scriptures/bofm/introduction

[4] 'Introduction' to the *Pearl of great price*. Retrieved from http://lds.org/scriptures

[5] 'Introduction' to the *King Follett Sermon*. Retrieved from http://www.lds.org/ensign/1971/04/the-king-follett-sermon and http://www.lds.org/ensign/1971/05/the-king-follett-sermon

[6] Smith Jr., J. *The King Follett Sermon*. Retrieved from https://www.lds.org/ensign/1971/04/the-king-follett-sermon

[7] Smith Jr., J. *The King Follett Sermon*.

[8] Bushman, p. 7.

Smith continues, explaining, "The head God called together the Gods and sat in grand council to bring forth the world. The grand councilors sat at the head in yonder heavens and contemplated the creation of the worlds which were created at the time...."[9] The plan they put into action was for the Gods to impose order on pre-existing chaos: "Let us go down. And they went down at the beginning, and they, that is the Gods, organized and formed the heavens and the earth" (Book of Abraham 4:1).

Because, according to Mormon belief, the elements are eternal, the Gods did not create the material universe but only imposed order on the constituent building blocks that already existed. In Smith's words, from his King Follett Sermon, "we infer that God had materials to organize the world out of chaos—chaotic matter, which is element, and in which dwells all the glory. Element had an existence from the time He had. The pure principles of element are principles which can never be destroyed; they may be organized and re-organized, but not destroyed. They had no beginning and can have no end..."[10]

In the Book of Moses, which Mormon followers believe was revealed to Joseph Smith, Jr. in June of 1830, Smith reports that he had a vision in which God speaks to Moses. "Moses lifted up his eyes unto heaven," (Moses 1:24) and the Lord tells him "And worlds without number have I created" (Moses 1:33). "And he beheld many lands; and each land was called earth, and there were inhabitants on the face thereof" (Moses 1:29). God, however, chooses to tell Moses only about the inhabitants of Moses's own planet: "But only an account of this earth, and the inhabitants thereof, give I unto you. For behold, there are many worlds that have passed away by the word of my power. And there are many that now stand, and innumerable are they unto man; but all things are numbered unto me, for they are mine and I know them" (Moses 1:35).

A few verses later, we learn that Enoch, a seventh generation descendant of Adam, has also learned that God has created not one Earth, but millions of earths. Enoch proclaims, "And were it possible that man could number the particles of the earth, yea, millions of earths like this, it would not be a beginning to the number of thy creations" (Moses 7:30).

When God created a universe full of earths, using the eternal elements as the material for his handiwork, matter turns out to be only one of the components of the eternally existing universe that predates the creative efforts of God. Intelligence, or "the soul—the mind of man —the immortal spirit,"[11] is also eternal: "Man was also in the beginning with God. Intelligence, or the light of truth, was not created or made" (Doctrines and Covenants 93:29). Whether the eternal intelligence includes the eternal souls of all individual persons as separate intelligences or is some kind of "great soup of intelligence"[12] is unclear. Whatever the answer, "The spirit of man ... does not have a beginning or end. ... God never had power to create the spirit of man [just as] God himself could not create himself. Intelligence is eternal and it is self-existing [sic]."[13]

---

[9] Smith Jr., J. *The King Follett Sermon.*

[10] Ibid.

[11] Ibid.

[12] Bushman, p. 72.

[13] Smith Jr., J. (1980). In A. F. Ehat & L. W. Cook (Eds.), *The words of Joseph Smith: The contemporary accounts of the Nauvoo discourses of the Prophet Joseph* (p. 346). Provo, UT: Religious Studies Center, Brigham Young University, as quoted in Bushman, p. 72.

In the Book of Abraham (3:19), God explains to Abraham that the spirit of the Lord is more highly placed than the primal human intelligences: "And the Lord said unto me: These two facts do exist, that there are two spirits, one being more intelligent than the other; there shall be another more intelligent than they; I am the Lord thy God, I am more intelligent than they all."

God then makes the decision to help the lesser souls achieve eternal life as gods, to become, in the language of Mormonism, exalted. Joseph Smith, Jr. explains God's decision making in this regard in his King Follett Sermon as follows: "God himself, finding he was in the midst of spirits and glory, because he was more intelligent, saw proper to institute laws whereby the rest could have a privilege to advance like himself. The relationship we have with God places us in a situation to advance in knowledge. He has power to institute laws to instruct the weaker intelligences, that they may be exalted with Himself."[14] As recorded in the Book of Moses (1:39), the Lord told Moses that this activity is God's greatest: "For behold, this is my work and my glory—to bring to pass the immortality and eternal life of man."

Joseph Smith, Jr., explains in the King Follett Sermon that in order to do this, God first creates human bodies in which he will place the human intelligences: "God made a tabernacle and put a spirit into it, and it became a living soul."[15] Smith further instructs his listeners that God's instructions for humans will save both our bodies and our spirits: "those revelations which will save our spirits will save our bodies. God reveals them to us in view of no eternal dissolution of the body, or tabernacle."[16]

Mormonism offers a way of life that shows believers how to follow those instructions given by God that Mormons believe will allow human spirits and bodies to achieve salvation, which according to Mormon belief is available to virtually all persons, but only to those persons, who accept Jesus Christ as Savior. As Joseph Smith, Jr. told those who attended the funeral of King Follett, "there is a salvation for all men, either in this world or the world to come, who have not committed the unpardonable sin. ... All sins shall be forgiven, except the sin against the Holy Ghost; for Jesus will save all except the sons of perdition. What must a man do to commit the unpardonable sin? ... he has got to deny Jesus Christ when the heavens have been opened unto him, and to deny the plan of salvation with his eyes open to the truth of it; and from that time he begins to be an enemy."[17]

To make sure these ideas from the King Follett Sermon are clear to those who are or seek to become Mormons, these ideas are restated in no uncertain terms in the scriptural Doctrines and Covenants: "They are they who are the sons of perdition, of whom I say that it had been better for them never to have been born; For they are vessels of wrath, doomed to suffer the wrath of God; with the devil and his angels in eternity; Concerning whom I have said there is no forgiveness in this world nor in the world to come—Having denied the Holy Spirit after having received it, and having denied the Only Begotten Son of the

---

[14] Smith Jr., J. *The King Follett Sermon.*

[15] Ibid.

[16] Ibid.

[17] Ibid.

Father, having crucified him unto themselves and put him to an open shame. These are they who shall go away into the lake of fire and brimstone, with the devil and his angels— And the only ones on whom the second death shall have any power; Yea, verily, the only ones who shall not be redeemed in the due time of the Lord, after the sufferings of his wrath. For all the rest shall be brought forth by the resurrection of the dead" (Doctrines and Covenants 76:32–39).

Those Mormons who achieve the highest level of salvation will be taken up into the celestial realm, where they will be equals with God and have their own worlds. Each such world will be occupied with living beings, intelligences, seeking their own salvation and immortality. In his 1930 work *A comprehensive history of the Church of Jesus Christ of Latter-day Saints: Century*, Brigham Henry Roberts (1857–1933 C. E.) wrote that "The Prophet Joseph Smith taught that these worlds and world-systems ... are or will be inhabited by sentient beings. This is assumed in all his revelations. It is everywhere taken for granted."[18]

But are these other worlds inhabited now? Mormon writings answer in the affirmative. In a vision given to Joseph Smith, Jr. on February 16, 1832, which is recorded in Doctrines and Covenants, Mormons are taught that these other beings are also their spiritual brothers, sisters and cousins, as "That by him, and through him, and of him, the worlds are and were created, and the inhabitants thereof are begotten sons and daughters unto God" (Doctrine & Covenants 76: 24). This vision, in Mormon scripture, clearly indicates the existence of a plurality of *inhabited* worlds. Furthermore, the God whom Mormons worship is the God of all these worlds and is responsible for populating all of these worlds.

And if currently-inhabited worlds do exist today, what is the status of those other created species in God's grand design for the universe? From Smith's divinely inspired vision that is preserved in the Book of Moses, we learn that God has both created and destroyed worlds: "The heavens, they are many, and they cannot be numbered unto man; but they are numbered unto me, for they are mine. And as one earth shall pass away, and the heavens thereof even so shall another come; and there is no end to my works, neither to my words" (Moses 1: 37–38).

To Enoch, the great-grandson of Adam and great-grandfather of Noah based on the stories recorded in *Genesis*, God reveals that while other species are not pure, humans on Earth are the most wicked creatures among all of God's created beings: "Behold, I am God; Man of Holiness is my name; Man of Counsel is my name; and Endless and Eternal is my name, also. Wherefore, I can stretch forth mine hands and hold all the creations which I have made; and mine eye can pierce them also, and among all the workmanship of mine hands there has not been so great wickedness as among thy brethren" (Moses 7:35–36). Thus, while other created beings may share with humanity the trait of wickedness, humans outdo them in this regard.

In the nineteenth century, some of Smith's followers were very clear in describing what some of these extraterrestrials might be like. Oliver B. Huntington (1823–1909 C. E.) first met Joseph Smith, Jr. in the Huntington family house in Watertown, New York in 1835. Huntington, who was ordained as an elder in the Latter-Day Saints in 1843, wrote

---

[18]Crowe, M. J. (1999). *The extraterrestrial life debate 1750–1900* (p. 242). Minneola, NY: Dover Publications.

in his autobiography that the man he called 'Joseph the Seer' could see "whatever he asked the Father in the name of Jesus to see." He proceeds to describe one of Smith's visions: "The inhabitants of the moon are more of a uniform size than the inhabitants of the earth, being about 6 ft in height. They dress very much like the quaker style and are quite general in style or the one fashion of dress. They live to be very old; coming generally, near a thousand years."[19]

The *Journal of Discourses* (a non-scriptural compilation of sermons and other materials from the early days of the Mormon Church) includes the text of a sermon given on July 23, 1870 by Brigham Young (1801–1877 C. E.), who served as President of the Latter-day Saints from 1847 until 1877. In this sermon, Brigham Young makes clear that he subscribes to the plurality of worlds hypothesis and the principle of plenitude, ascribing inhabitants to both the Moon and the Sun: "Who can tell us of the inhabitants of this little planet that shines of an evening, called the moon? When we view its face we may see what is termed 'the man in the moon,' and what some philosophers declare are the shadows of mountains. But these sayings are very vague, and amount to nothing; and when you inquire about the inhabitants of that sphere you find that the most learned are as ignorant in regard to them as the most ignorant of their fellows. So it is with regard to the inhabitants of the sun. Do you think it is inhabited? I rather think it is. Do you think there is any life there? No question of it; it was not made in vain. It was made to give light to those who dwell upon it, and to other planets; and so will this earth when it is celestialized."[20] Such ideas are very much in keeping with intellectual thought of the eighteenth and early nineteenth centuries, though by the late nineteenth century, when Brigham Young offered this sermon, only a few professional astronomers continued to think that either the Moon or Sun or any other part of our solar system was populated.

A century later, Mormon leaders no longer speak often about inhabitants of the Sun and Moon, but the concept of human-like extraterrestrials remains deeply ingrained in Mormon belief. In 1970, Joseph Fielding Smith (1876–1972 C. E.), the tenth President (1970–1972) of the Latter-day Saints Church, wrote, "We are not the only people that the Lord has created. We have brothers and sisters on other earths. They look like us because they, too, are the children of God and were created in his image, for they are also his offspring."[21] Neal A. Maxwell (1926–2004 C. E.), a Latter-day Saints apostle (Latter-day Saints apostles are "special witnesses of Jesus Christ, called to teach and testify of Him throughout the world"[22]) who served as a member of the Quorum of the Twelve Apostles (the second-highest governing body, after the First Presidency, in the Church of Latter-day Saints) from 1981 until 2004, echoed this same idea in 1990, writing, "We do not know how many inhabited worlds there are, or where they are. But certainly we are not alone."[23]

---

[19] Huntington, O. B. *History of the life of Oliver B. Huntington, Written by Himself, 1878–1900* (p. 10), typescript manuscript. Retrieved from http://www.archive.org/stream/historyoflifeofo00hunt#page/n3/mode/2up

[20] *Journal of Discourses, 13*, 271. Retrieved from http://jod.mrm.org/13/268#271

[21] Smith, J. F., & McConkie, B. R. (Ed.), *Doctrines of salvation. Sermons & writings of Joseph Fielding Smith* (vol. 1, p. 62).

[22] http://www.lds.org/church/leaders/quorum-of-the-twelve-apostles

[23] Maxwell, N. A. (1990). *Wonderful flood of light* (p. 25). Salt Lake City: Bookcraft Publications.

We can assume that Mormons would embrace, without any concerns, any future announcement made by astronomers regarding the discovery of extraterrestrial life. Even were astronomers able to offer proof that life is absent on all planets thus far surveyed, Mormons could also easily accept that knowledge. Any such evidence offered for the absence of intelligent life elsewhere in the universe could be understood as indicating that those sentient beings live outside of the limits of our surveys, rather than that they do not exist.

Of equal importance, for Mormons life *on Earth* is absolutely central and critical to Mormon belief. Life on Earth is a brief interlude for the eternal spirits of Mormon believers. It is also a gift from God that offers each intelligence the opportunity "to choose godliness over worldliness"[24] and become exalted. Those living on other worlds either have not yet been born into an earthly existence or have already experienced life in human form on Earth. This assertion for the specialness of Earth and life on Earth means that Mormonism is spiritually and theologically Earth-centered. Intelligences are given bodies on this planet; they are given free will while on this planet to make choices that will affect their condition for all of eternity; and Jesus lived and gave his life on this planet, not on any other planet, so that believers might be redeemed by accepting him as their Savior.

Given the theological geocentrism in Mormonism, we might ask how Mormons would react to life forms on other planets that are not made in God's humanlike image. If these life forms are deemed lesser beings, without rationality and thus without intelligences, they would have no impact on Mormon thought; however, if these beings are intelligent, could they be Mormons? If not, could or should Mormons try to convert them to Mormonism? As converts, would they have to spend time in physical form on Earth? Or could a convert exercise free will, make appropriate choices between godliness and worldliness and, while a resident on another world, accept or reject the salvation offered by Jesus?

While Mormonism appears to be ready to fully embrace extraterrestrial life, whether those extraterrestrials are in human form or are easily labeled as lesser beings, Mormonism nevertheless will have some theological issues with which to wrestle. That theological wrestling match may recently have become easier for Mormons to manage, as in late 2013 Dieter F. Uchtdort, one of the three persons in the 'First Presidency' of the Church of Latter-day Saints, and thus one of two top counselors who advise current Mormon president and prophet Thomas S. Monson, said "there are times when members or leaders in the church have simply made mistakes." Mormon scholars Terryl and Fiona Givens "said that Mr. Uchtdorf's talk was seminal because of the admission that church leaders were fallible."[25] Such a position opens the door for recognizing, for example, that nineteenth-century claims by church leaders, now understood to be factually erroneous, about the reality of living beings on the Moon and the Sun were the result of influences by then-popular cultural ideas and were not divinely-inspired revelations. As of today, however, no such revisions in Mormon theology have been put forward by today's leaders.

---

[24] Bushman, p. 75.

[25] Goodstein, L. (2013, October 8). A leader's admission of 'Mistakes' heartens some doubting Mormons. *New York Times*.

15

Islam

Whatever beings there are in the heavens and the earth do prostrate themselves to Allah.

Qur'an (13:15)

Islam, like other faiths, has fundamentalist and conservative traditions. All Muslims, however, likely would agree that the prophetically revealed religion of Islam is a set of practices designed only for humans on Earth, so that they can submit to the will of Allah, who is God of the entire universe. The discovery of extraterrestrials should only strengthen a Muslim's faith in the presumed infinite power of God.

"The most common attitude toward science in the Islamic world is to see it as an objective study of the world of nature, namely as a way of deciphering the signs of God in the cosmic book of the universe. Natural sciences discover the Divine codes built into the cosmos by its Creator, and in doing so, help the believer marvel at the wonders of God's creation."[1] This view of the relationship between Islam and science, as expressed by George Washington University scholar of Islamic philosophy Ibrahim Kalin, is echoed by Stanford University professor of Middle Eastern History Ahmad Dallal, who has written, "Almost all sources agree that the Qur'an condones, even encourages, the acquisition of science and scientific knowledge and urges people to reflect on the natural phenomena as signs of God's creation."[2]

---

[1] Kalin, I. (2002). Three views of science in the Islamic world. In T. Peters, M. Iqbal, & S. N. Haq (Eds.), *God, life, and the cosmos: Christian and Islamic perspectives* (p. 47). Burlington, VT: Ashgate Publishing.
[2] Dallal, A. (2010). *Islam, science, and the challenge of history* (p. 117). New Haven: Yale University Press.

D.A. Weintraub, *Religions and Extraterrestrial Life: How Will We Deal With It?*, Springer Praxis Books, DOI 10.1007/978-3-319-05056-0_15, © Springer International Publishing Switzerland 2014

Given that "the predominant Islamic perspective sees reason and revelation as harmonious and complementary,"[3] any discovery of extraterrestrial life by scientists likely would be embraced by Muslims. Islamic sacred scripture, however, at least as understood by virtually all Islamic scholars now and in the past, goes further, as the Qur'an appears to claim that spiritual beings beyond the Earth do exist. If this is so, we need to ask whether these extraterrestrials could be Muslims and if they would follow the faith of Islam according to the tenets of the faith laid out for them by the prophet Muhammad. We also should ask whether one can be a Muslim without being a follower of the Islamic faith.

The religion of Islam arose about 1,400 years ago when, according to Islamic belief, in the year 610 C. E., the archangel of revelation Gabriel revealed the word of God to the prophet Muhammad. About a decade later, in 622 C. E., Muhammad led his followers across the Arabian Peninsula from Mecca to Medina. Within a century after Mohammad's death in 632 C. E. the Rightly Guided Caliphs, who led the Muslim community and the Arab armies in support of Islam, quickly spread the Islamic faith as far east as Iran and as far west as Spain. In today's world, Islam is a global religion, with about 25 % of the population of the world (more than 1.5 billion people) identifying as Muslims.

Islam accepts as foundational texts the revelations of the prophets Moses, David, Jesus and Muhammad; Muslims believe, however, that the preserved forms of the Jewish scriptures and the Christian New and Old Testaments are corrupted by mistakes in transmission and translation; thus, Muslims revere the newer scripture, the *Qur'an*, the words of which, according to those of the Islamic faith, were revealed to Muhammad and written down by his followers. Muslims also hold as sacred a secondary text, the *Hadith*. Both sacred texts are believed to be perfectly recorded and preserved. Muslims believe that the Qur'an preserves "the verbatim revelation of God's word"[4] as revealed to Muhammad by the angel Gabriel, while the Hadith records the words and deeds of Muhammad, his family, and his immediate followers.

In order to understand what the writings, teachings and beliefs of Islam say about the universe and about life in the universe, and in order to ask whether spiritual extraterrestrial life forms could be Muslims, we can start with the six articles of the Islamic faith. A person must accept these articles of faith in order to be a member of this religious community. These six articles of faith are the belief in one God—God has no parents, children, partners, or siblings; having faith in the prophets, including, among others, Adam, Noah, Abraham, Moses, Jesus and Muhammad; the belief in God's invisible, spiritual beings, the angels; having faith in the Holy Scriptures (the Torah revealed to Moses, the Psalms revealed to David, the Gospel revealed to Jesus, the Qur'an revealed to Muhammad); the belief in God's plan and will, that God has knowledge of all that will happen; and the belief in the Day of Judgment, when all souls will be divided between those who will and those who will not enter Paradise.

---

[3] Chittick, W. C. (2002). The anthropocosmic vision in Islamic Thought. In T. Peters, M. Iqbal, & S. N. Haq (Eds.), *God, life, and the cosmos: Christian and Islamic perspectives* (p. 127). Burlington, VT: Ashgate Publishing.

[4] Nasr, S. H. (2007). In W. C. Chittick (Ed.), *The essential Seyyed Hossein Nasr* (p. 57). Bloomington: World Wisdom.

In addition to accepting the six articles of faith, those who identify as Muslims accept as religious obligations a number of duties identified as the five pillars of the faith. These five pillars are professing one's faith; praying five times a day while facing in the direction of Mecca; paying an alms tax; fasting during daylight hours during the month of Ramadan; and undertaking a pilgrimage to Mecca.

George Washington University professor Seyyed Hossein Nasr, an influential modern Islamic historian and philosopher, asserts, "All beings in the universe, to begin with, are Muslim, i.e., 'surrendered to the Divine Will.'"[5] Nasr's view is hard to reconcile with the actual practices of Islam unless being a Muslim is not identical with practicing the Islamic faith. One might plausibly conclude from two of the pillars of the faith (facing Mecca during prayer; a pilgrimage to Mecca) that Islam is an Earth-centered religion that can only be practiced by an Earth-bound being born after 610 C. E., since Mecca, a city on Earth that itself is no more than a few thousand years old, appears to be of central importance for two of the five pillars of the faith. If extraterrestrials exist and live at a great remove from Earth, how would they determine the direction of Mecca, five times every day, for prayer? And how would they travel to Mecca, possibly traversing thousands or millions of light years, during lives that most likely are much too short to survive such an interstellar or intergalactic voyage? Taken further, have extraterrestrials been visiting Mecca regularly for the last 1,400 years, and could humans or extraterrestrials have been practicing members of the Islamic faith prior to the revelation of the Qur'an to Muhammad, or before the holy city of Mecca emerged from the sands of the Arabian Peninsula?

The five pillars of faith offer further complications for extraterrestrials. Planets around other stars need not spin at the same rate as the Earth. Thus, a day on a quickly-spinning planet orbiting a star other than the Sun might be as short as only a few earth hours, while a slowly-spinning planet might have a day as long as a few earth years. If supplicants still must pray five times a day, even on a world with a "day" as short as 3 h, would they also have time to eat, work, play or sleep? If they prayed five times a day on a world with a 10,000-h "day," would such behavior offer sufficient evidence of the supplicant's submission to the will of God?

The position advocated by Nasr, that extraterrestrials can or do exist and can be and are Muslims, appears only plausible if being a Muslim and being a member of the Islamic faith are not always identical: science and logic together do not allow one to practice Islam elsewhere in the universe, as a faithful being in a galaxy far, far away could not journey to Mecca or pray toward Mecca, could not follow the pillars of the faith and therefore could not be a follower of Muhammad. Pakistani physicist Pervez Hoodbhoy describes Nasr's "brand of Islamic orthodoxy" as being anti-science, because science "relies solely on reason and observation as the arbiters of truth." The scientific approach to knowledge, Hoodbhoy says, is unacceptable to Nasr.[6] In a sense, though, Nasr is being logically consistent, as he seeks his answers through revelation rather than via the logic of science.

---

[5] Nasr, S. H. (2001). *Science and civilization in Islam* (p. 23). Retrieved from http://hdl.handle.net/2027/heb.03330.0001.001

[6] Hoodbhoy, P. (1991). *Islam and science: Religious orthodoxy and the battle for rationality* (p. 70). London: Zed Books.

In addition, a middle position does exist, one in which being a follower of Islam and being a Muslim are not one and the same.

If Islamic scholarship can provide a basis for supporting the idea that extraterrestrials do exist, we could conclude that the Islamic religious duties associated with locating and visiting Mecca and related to the uniqueness of the Earth (the 24-h day) are of importance in guiding religious practice only for Muslims on Earth. Extraterrestrial Muslims could demonstrate their faith according to local circumstances that vary with one's location in the universe. One could, in fact, distinguish between being Muslim and being a member of the Earth-specific, Earth-only Islamic faith. Perhaps this is the meaning of Nasr's words.

Every other world populated with sentient beings could, for example, have its own prophet, its own Qur'an-like revelation and its own pillars of the faith. Those beings would be Muslims if a Muslim is understood as one who is a sentient being who surrenders to the Divine Will; but, extraterrestrials would not be followers of the Islamic faith. Islam would then be defined by the practices of Muslims on Earth, who would be the only such beings for whom the particular revelation known as the Qur'an and the wisdom of the prophet Muhammad applies. Muslims in other parts of the universe would have their own prophet, their own revelation and thus their own religion.

We can step back from the question as to whether all extraterrestrial beings could be Muslims and ask, first, in the view of Muslims on Earth, is Allah the God of the entire universe? The Islamic answer is yes. Mehdi Golshani, Chairman of the Faculty of the Philosophy of Science at Sharif University of Technology in Tehran, writes, "According to the Qur'an, everything is created by God."[7] Several Qur'anic verses oft-cited in support of this statement of the Universal God, are "Allah is the Creator of all things: He is the One, the Supreme and Irresistible" (Qur'an 13:16) and "Allah is Creator of all things, and He is in Guardian over all things." (Qur'an 39:62).[8] The most important verse in this regard, and one that is also one of the most important verses in the Qur'an, is likely the one known as the Throne verse, which says, in part, "His are all things in the heavens and on the earth. … His throne doth extend over the heavens and the earth." (Qur'an 2:255)

In the words of Nasr, not only does God's dominion extend over the entire universe, everything that exists, whether in this or any other universe, is created by God. Furthermore, everything that possibly could be created by an Infinite God has been created. In writing about the verse "So glory to Him in Whose hands is the dominion of all things" (Qur'an 36:83), Nasr writes, "God as Reality is at once absolute, infinite, and good or perfect. … Ultimate Reality contains the source of all cosmic possibilities … with God all things are possible."[9]

The theological opinion that creating other worlds is well within the realm of the many cosmic possibilities for a God for whom all things are possible and who has domin-

[7] Golshani, M. (2002). Creation in the Islamic outlook and in modern cosmology. In T. Peters, M. Iqbal, & S. N. Haq (Eds.), *God, life, and the cosmos: Christian and Islamic perspectives* (pp. 223–224). Burlington, VT: Ashgate Publishing.

[8] All Qur'anic translations by Yusuf Ali. Retrieved from http://corpus.quran.com/translation.jsp

[9] Nasr, S. H. (2007). In W. C. Chittick (Ed.), *The essential Seyyed Hossein Nasr* (p. 155). Bloomington: World Wisdom.

ion over all that is has been in the mainstream of Islamic thought for a thousand years. The twelfth-century philosopher Fakhr al-Din al-Razi (1149–1209 C. E.), who is celebrated even today for his commentary on the Qur'an known as *The Keys to the Unknown*, argued that Aristotle's ancient and powerful logical argument concerning the uniqueness of our world does not stand up to Qur'anic scrutiny: "It has been proven by evidence that God, the Exalted, is capable of actualizing all possibilities. Thus, be He Exalted is capable of creating thousands and thousands of worlds beyond this world, each of which would be greater and more massive than this world ... and the argument of the philosophers for the uniqueness of this world is weak and poor, being based on invalid premises."[10]

Seen in this light, several verses in the Qur'an that seem to imply the non-uniqueness of the Earth and the existence of other worlds can be understood more clearly as affirmations for believers of the existence of those worlds. Among these verses are "Praise be to Allah, the Cherisher and Sustainer of the worlds" (Qur'an 1:2) and "He is the Lord of (all) the Worlds" (Qur'an 41:09). Another verse suggests the existence of perhaps seven Earth-like worlds—"Allah is He Who created seven Firmaments and of the earth a similar number" (Qur'an 65:12)—while another suggests that we can never know how many such worlds might exist—"And He has created (other) things of which ye have no knowledge" (Qur'an 16:8).

Given that the starting point for understanding the Qur'an for Muslims is that the words therein are the verbatim words of God as revealed to Muhammad, these phrases offer clear evidence to Muslims that the universe has multiple worlds to which Allah pays careful attention. But are they populated? While none of these verses say anything explicit about the existence of life on those worlds, one might wonder why Allah would be interested in sustaining, let alone cherishing, lifeless worlds, and therefore one might surmise that these other worlds are almost certainly inhabited.

The Qur'an does not leave the faithful guessing as to whether extraterrestrial life exists. As early as the first century after the birth of Islam, one of the most revered of the early leaders of Islam, Imam Muhammad al-Baqir (676–733 C. E.), wrote, "Maybe you see that God created only this single world and that God did not create Homo sapiens besides you. Well, I swear by God that God created thousands and thousands of worlds and thousands and thousands of humankind."[11] The conclusion that these other worlds are populated with beings capable of submitting to the Divine Will, i.e., to being Muslims, is supported by several verses in the Qur'an. Based on the verse (42:29)—"And among His Signs is the creation of the heavens and the earth, and the living creatures that He has scattered through them: and He has power to gather them together when He wills"—many Islamic scholars strongly suggest that living creatures exist beyond the Earth. Mirza Tahir

---

[10] Fakhr al-Din al-Razi, quoted in Golshani, M. (2002). Creation in the Islamic outlook and in modern cosmology. In T. Peters, M. Iqbal, & S. N. Haq (Eds.), *God, life, and the cosmos: Christian and Islamic perspectives* (p. 225). Burlington, VT: Ashgate Publishing.

[11] Imam Muhammad al-Baqir, quoted in Golshani, M.(2002) Creation in the Islamic outlook and in modern cosmology. In T. Peters, M. Iqbal, & S. N. Haq (Eds.), *God, life, and the cosmos: Christian and Islamic perspectives* (p. 225). Burlington, VT: Ashgate Publishing.

Ahmad, who was Khalifatul Masih IV, Caliph of the Ahmadiyya Muslim Community in Pakistan before he was exiled in 1984 (the Pakistani National Assembly declared this group 'non-Muslim') asserts that "The second part of the same verse speaks not only of the possibility of extraterrestrial life, but it categorically declares that it does exist."[12]

Like humans on Earth, these beings belong to Allah—"To Him belong all (creatures) in the heavens and on earth" (Qur'an 21:19). They also pray to Allah—"Whatever beings there are in the heavens and the earth do prostrate themselves to Allah" (Qur'an 13:15). In addition, like human members of the Islamic faith, they will be judged by Allah—"Not one of the beings in the heavens and the earth but must come to (Allah) Most Gracious as a servant. He does take an account of them (all), and hath numbered them (all) exactly. And every one of them will come to Him singly on the Day of Judgment" (Qur'an 19:93–95).

Islamic scripture appears to make very clear that whatever these beings in the heavens may look like, they are rational, intelligent life forms who are capable of praying to Allah and will be held accountable before Allah on Judgment Day. Given that one of the articles of the Islamic faith is to believe in invisible spiritual beings known as angels, one possibility is that these extraterrestrial beings are angels, and, since angels are invisible, astronomers might never discover evidence of their presence in astronomical studies.

Could these beings be angels? And would angels need to prostrate themselves before God? Again, the Qur'an answers these questions. No, they are not angels. Ahmad explains that the word translated in verse 42:29 as 'living creatures' (da'bbah) refers to living, breathing, walking organisms and not to angels. "Da'bbah covers all animals which creep or move along the surface of the earth. It does not apply to animals which fly or swim. It is certainly not applicable to any form of spiritual life. In Arabic a ghost will never be referred to as da'bbah, nor an angel for that matter."[13] The word used in verse 42:29 for 'together' (jam-'i-him) is, according to Ahmad, an Arabic expression "which specifically speaks of bringing together of life on earth and the life elsewhere."

Ahmad's interpretation is widespread among Islamic scholars. Muhammad Asad, a prominent modern Muslim scholar, writes, similarly, "The word da'bbah denotes any sentient, corporeal being capable of spontaneous movement; it is contrasted here with the non-corporeal, spiritual beings designated as angels."[14] Verse 16:49 of the Qur'an, in which da'bbah (the moving creatures) are contrasted with the angels, makes this notion explicit: "And to Allah doth obeisance all that is in the heavens and on earth, whether moving (living) creatures or the angels: for none are arrogant (before their Lord)." Abdullah Yusuf Ali, an Indian Islamic scholar whose English language translation of the Qur'an is in widespread use, writes, in regard to the verse 42:29, "It is reasonable to suppose that

[12] Ahma, M. T. 'Part IV: The Quran and extraterrestrial life' from *Revelation, rationality, knowledge & truth*. Retrieved July 11, 2012, from http://www.alislam.org/library/books/revelation/part_4_section_7.html

[13] Ibid.

[14] Asad, M. (2003). *The message of the Qur'an* (p. 449). England: Book Foundation, as quoted in http://www.onislam.net/english/ask-about-islam/islam-and-the-world/worldview/167124-aliens-and-extraterrestrial-life-an-islamic-look.html; accessed July 12, 2012.

Life in some form or another is scattered through some of the millions of heavenly bodies scattered through space."[15]

No doubt, the existence of extraterrestrial life is compatible with Islam. The Qur'an appears to tell believers that intelligent living beings who are not angels and who worship Allah live in places other than on Earth. But would these extraterrestrials be followers of the religion known as Islam, or are they only Muslims? To determine whether Islam is a universal religion requires that we decide what criteria determine whether one is a Muslim.

Because none of the six articles of faith are Earth-centric, if one must only accept the six articles of the faith in order to be a Muslim, then any and all extraterrestrials could be Muslims. These extraterrestrial spiritual beings could pray to Allah according to the practices established for them by a prophet who arose in their own world. They could be Muslims but not followers of Muhammed, not practicers of the Islamic faith.

If one must also follow the five pillars of faith in order to be a Muslim, then all Muslims must live on Earth, or at least in the practical local vicinity of Earth (one can imagine a future resident of Mars praying toward Mecca and undertaking a pilgrimage to Mecca). If so, then the religion we call Islam would be confined to the Earth. The religion of Islam would provide a set of behavioral rules that brings Earth-bound worshipers into a relationship with Allah. In this case extraterrestrials would not be Muslims. Such a conclusion appears inconsistent with a millennium of Islamic theological development. The geocentric practices of Islam, including the requirement that one face Mecca during the five daily prayers and that one should undertake a pilgrimage to Mecca, can be understood as historical and appropriate only for Earth-bound worshippers of Allah and not fundamental to the worship of Allah for all beings in all locations in the universe.

Some Islamic scholars argue that the Qur'an advocates praying in the direction of Mecca in order to equip believers with a physical pointer toward the presence of Allah; doing so also provides unity to the worldwide (on Earth) community of Muslims. Similarly, praying five times every day imposes a firm but manageable expectation on practicing Muslims on Earth, given the approximate 16 h during which most persons are awake during a terrestrial day and the amount of time people need each day while awake for working, eating, playing and taking care of hygiene. Prayers to Allah, however, are not contingent on these practices, and one could easily imagine spiritually inclined beings beyond the Earth finding other ways to place themselves in the presence of Allah and different ways to submit themselves to His will.

Were it possible to conclude otherwise, that life on Earth is unique in the universe, Islamic theologians would have to reinterpret some scriptural passages; however, Islamic belief for human followers of the Islamic faith would be unaffected as it does not in any way depend on, succeed or fail on the basis of the existence of extraterrestrial life.

Nasr blurs the lines between those who follow the practices of a terrestrial religion and the religion itself. "If a religion were to cease to exist on earth," he writes, "that does not mean that it would cease to possess any reality whatsoever. In this case, its life cycle on earth would have simply come to an end, while the religion itself ... would subsist in

---

[15] Yusuf Ali, A. (1938). *The Qur'an: Text, translation and commentary* (p. 1314). Beirut: Ad-Dar Al-'Arabiah, as quoted in http://www.onislam.net/english/ask-about-islam/islam-and-the-world/worldview/167124-aliens-and-extraterrestrial-life-an-islamic-look.html. Retrieved July 12, 2012.

the Divine Intellect in its trans-historical reality. ... the archetypal reality which the religion represents would persist." He continues, "religion itself cannot be reduced to its terrestrial embodiment. If a day would come when not a single Muslim or Christian were to be left on the surface of the earth, Islam and Christianity would not cease to exist nor lose their reality in the ultimate sense."[16] In Nasr's metaphysics, all "authentic religions come from the same Origin," and so even though "from the operative and practical point of view ... at a particular juncture in history," some religions may "decay and even die in the sense that their earthly careers terminate," the Ultimate Reality of religion would continue to exist.[17] And in Nasr's Ultimate Reality, Islam has a reality that is bigger than the Earth, bigger than the sum of all the actions and beliefs of all Muslims on Earth.

Ultimately, in Nasr's view the Islamic faith is defined solely by just two of the articles of faith. "At the heart of Islam stands the reality of God, the one, the Absolute and the Infinite ... 'There is no god but God' is the first of two testifications (*shahadah*) by which a person bears witness to being a Muslim; the second is *Muhammadun rasul Allah*, 'Muhammad is the messenger of God.'"[18] And so his answer seems clear: one is a Muslim if one surrenders oneself to the One God, and one is a member of the Islamic faith (the prophetically revealed religion founded by Muhammed) if one is a follower of Muhammad and accepts the articles of the Islamic faith and puts into practice the pillars of the Islamic faith.

Islam, according to Nasr's logic, is for humans on Earth and only for humans on Earth. Evidence for the existence of extraterrestrial life would be yet another page in the vast book of the cosmos, written by Allah, the Author. In reading that page, followers of the Islamic faith would learn more about God, and in doing so would strengthen their faith, as the God of Islam intends for human beings to read the book as written.

---

[16] Nasr, S. H. (2007). In W. C. Chittick (Ed.), *The essential Seyyed Hossein Nasr* (p. 22). Bloomington: World Wisdom.

[17] Ibid (p. 27).

[18] Ibid (p. 43).

16

Hinduism

In the higher planets of the material world, the yogis can enjoy more comfortable and more pleasant lives for hundreds of thousands of years.

C. Bhaktivedanta Swami Prabhupada

Hinduism is a descriptive term coined by Europeans in the eighteenth and nineteenth centuries. It places a large family of belief systems with similar and overlapping practices and intellectual philosophies under a single religious umbrella. In the early twenty-first century about a billion people, most of whom live on the Indian subcontinent, can be found under the Hindu umbrella. Hinduism has no single founder or prophet, no Abraham or Buddha or Jesus or Muhammad; Hindus do not all practice the same rites or say the same prayers; meditation and mysticism are sufficient for some Hindus while other Hindus worship gods, hold ceremonies and follow traditions as important aspects of their spirituality; and though voluminous important Hindu scriptures exist, including the *Vedas* (which date back to the period from 1500 to 1000 B. C. E.), the *Upanishads* (dating to about 1200–500 B. C. E.), and the *Bhagavad Gita* (composed between the fifth and second centuries B. C. E.), Hindus do not consider any of these texts as containing infallible truths or words dictated by a god to a human prophet. Instead, Hindu scriptures provide guidance and wisdom from previous generations of Hindus who have sought and perhaps found enlightenment about the universe.

Several important and powerful ideas in spiritual thinking do serve as unifying principles for Hinduism. First, for Hindus, *Brahman* is the formless and eternal universal spirit that is the origin of the universe. Second, *atman*, the Sanskrit word meaning 'self' or 'breath,' is the imperceptible, incomprehensible, eternal and non-material living essence that transcends the physical body of a living being and connects each being with Brahman. Atman is not mind or consciousness and is similar to but not quite identical with the

D.A. Weintraub, *Religions and Extraterrestrial Life: How Will We Deal With It?*,
Springer Praxis Books, DOI 10.1007/978-3-319-05056-0_16,
© Springer International Publishing Switzerland 2014

western concept of the soul. Third, the two related concepts of *karma* and *samsara* are common to all Hindu belief systems. Understanding these concepts is fundamental for understanding how or if Hindu beliefs relate to the possibility of extraterrestrial life.

Karma comprises all the actions one undertakes during one's lifetime and all the consequences of those actions. Together, all these actions and consequences create the total ethical quality of one's life. Karma is impersonal. It is not administered or judged by a deity. It merely exists in and of itself as a stored cosmic force. And like a compressed spring, "the stored force of karma will manifest itself at some time or other."[1]

Samsara is the endless cycle of birth and rebirth within and between levels of life. Seen from the Hindu perspective of samsara, death is not a final end of life. Instead, death represents a moment during reincarnation when the individual soul, or *jiva*, of a living being, along with the spiritual essence underlying the existence of that jiva, transitions from one life in one material body to the next life in a new body, when one's jiva transmigrates to a new life or, if one escapes samsara, is released to fuse with Brahman. Reincarnation is almost guaranteed to transfer one's life force from the lowest level of the plants upward through the increasingly complex levels of animals until a jiva is reincarnated in human form. Huston Smith writes, in *The World's Religions*, "With the soul's graduation into a human body, this automatic, escalator-like mode of ascent comes to an end; its entry into this exalted habitation is evidence that the soul has reached self-consciousness, and with this estate comes freedom, responsibility, and effort."[2] Once reborn as a human, future rebirths enable the jiva to transmigrate further upwards to even higher levels of humans, including into any of the five recognized human castes. These castes include the *untouchables* at the bottom (but above the highest of the animals), then, moving upwards, the *shudras* (servants), further upwards the *vaishyas* (artisans, merchants, farmers and peasants), then the *kshatriyas* (warriors, rulers and administrators), and finally, at the top, the *brahmins* (intellectual and spiritual leaders). During samsara, a jiva can even transmigrate upward from one of the levels of humans to the level of the gods.

One's karma will determine the condition of one's rebirth. Because in Hindu belief animals and plants lack the ability to act rationally, they can only generate small amounts of good or bad karma during their lives. As a result, and because a Hindu rebirth at any level above the animals requires a significant amount of good karma, most rebirths will be as plants or insects. Humans, however, because they have the ability to freely make decisions, can generate large amounts of karma, both good and bad, during a single lifetime. As a result, humans have a great deal of control over their own destinies through the choices and decisions made by each individual during his or her lifetime. Thus, rebirth in human form offers a special opportunity for Hindus and is assumed to be a rare, almost miraculous event.

Good karma leads to favorable conditions upon rebirth. With a favorable enough reservoir of karma, one could be reborn as a god. On the other hand, bad karma leads to unfavorable conditions upon rebirth. In such a situation, one might be reborn as a slave, an animal or even a plant. Since, according to Hindu beliefs, karma affects gods as well as humans, even a god could be reborn into a lesser state if that god used up its good karma and did not escape samsara.

---

[1] Karbhari, B. F. (1913). *The karma philosophy* (p. 3). Madras: Brahmavdin Press.
[2] Smith, H. (1991). *The world's religions* (p. 64). New York: Harper One.

For Hindus, the existence of individuals is an illusion. We are all part of the eternal whole of Brahman, the absolute. Brahman is beyond the material but is also the building material of all reality. Brahman creates not by willing creation but simply as the natural result of Brahman's being. Brahman is beyond knowledge, words and images; Brahman transcends all human categories of description; Brahman is without qualities and, for humans, is inconceivable. All Hindu deities are manifestations of Brahman and they all are capable of mediating a spiritual path to Brahman.

The human ego prevents people from understanding Brahman fully, and so humans are trapped in samsara. Samsara is not desirable; therefore, the goal for Hindus is to achieve liberation from samsara, to end this cycle of birth and rebirth. How to escape from samsara is the existential problem at the heart of Hinduism (and Buddhism and Jainism). Escape from samsara is possible, but very difficult. By practicing Hinduism, one learns how to separate from one's egoistic self in order to achieve union with Brahman and thereby escape samsara.

The Upanishads are the earliest known historical documents that offer a solution to the existential problem of escaping samsara.[3] The Upanishad-based method of seeking union with Brahman through mystical knowledge, the so-called Path of Knowledge, is one of the three principal Hindu paths for achieving liberation from samsara. This path involves asceticism and spirituality; meditation is critical for transcending one's self via the Path of Knowledge.

During the time period from 300 to 220 B. C. E., Hindus developed an alternative, less abstract, more traditionally religious form of spirituality involving temples, pilgrimages and images of deities. In this form of Hinduism, which offers two additional paths for liberation from samsara, Hindus seek union with Brahman through devotion to and worship of a god who is a manifestation of Brahman. The first of these two additional paths is the Path of Action or Works. This path involves ritual worship of lesser deities and following social rules that are appropriate for one's caste. The second of these two additional paths is the Path of Devotion. This path is a life spent in service to the lord, one in which a Hindu will center his or her life on love of the principle deities rather than on oneself. A Hindu following this path would worship in temples.

According to the teachings of the *Bhagavad Gita*, which appeared at about this same time, devotion to a god, which is possible for all humans of all castes, can lead to liberation from samsara. One needn't practice mysticism or be an ascetic. One only needs to love and imitate Lord Krishna (the creator) and worship the god Vishnu (the preserver and protector of everything created).

Hindu ideas about the possibility of extraterrestrial life are inextricably connected to Hindu beliefs about samsara through Hindu cultural ideas about the structure of the universe. In Hindu cosmology, three worlds exist. The first is the physical universe inhabited by plants, animals, humans and all other possible kinds of living beings that are neither gods nor angels; the second is the mental plane of existence in which deities and angels live; the third is a spiritual universe where only Hindu gods reside. If extraterrestrials exist where humans might learn of their presence through signs of the physical impact of the Others on the worlds they inhabit, they must necessarily be living somewhere in the physical universe.

---

[3] http://www.sacred-texts.com/hin/sbe01/index.htm#contents

The Hindu belief in reincarnation places no limits on where in the physical universe a reincarnation event might take place. Furthermore, the laws of physics as understood by humans do not encumber the process of transmigration. Hindus do, however, expect that karma and samsara are universal, affecting any living being anywhere in the universe. A living being from Earth might therefore be reincarnated into another kind of living being on another planet anywhere in the universe, and those beings in turn could be reincarnated here on Earth. Furthermore, the generation of karma, good or bad, is not limited to the lives of beings on planet Earth. Extraterrestrials would be other beings who, like us, must struggle to increase their karma in order to escape samsara.

Hindus would not be at all surprised by the discovery of extraterrestrial beings. They also would not be surprised if astronomers fail to identify intelligent life in a survey that studies only a part of the Milky Way or even the entire galaxy. No matter how many millions of planets are surveyed, a Hindu could justifiably hold the opinion that such a survey would be small in both space and time in comparison to the Hindu view of the size and age of the universe.

An unknown for Hindus would be how those extraterrestrials fit into the hierarchical structure of plants, animals, humans and gods, because the Hindu hierarchy of living beings does not, at present, include extraterrestrials. If the extraterrestrial life forms we might find appear subhuman, immobile and most similar to plants on Earth, Hindus likely would classify them as plants. If, instead, the extraterrestrial life forms are mobile but are found to be arational, spiritless or mechanistic beings, they likely would be classified with the non-human animals. Sentient extraterrestrials, ones who appear to understand life and death, who can attempt to ask questions about eternity and infinity and who recognize their own spiritual natures, might be understood to have the same kind of freedom of choice to affect their karma as do humans. As such, they would be seen to occupy one or more of the same levels of living beings occupied by humans. In such a case, perhaps humans and extraterrestrials would be interchangeable, at least in the sense of transmigration and reincarnation. Finally, if an extraterrestrial civilization is sufficiently advanced so as to satisfy Arthur C. Clarke's third law—"any sufficiently advanced technology is indistinguishable from magic,"[4] Hindus might identify those beings as deities.

C. Bhaktivedanta Swami Prabhupada (1896–1977 C. E.)[5] was a prolific and authoritative translator of Indian religious writings and the founder of the International Society for Krishna Consciousness (more commonly known as the Hare Krishnas). Prabhupada wrote a commentary on *Srimad Bhagavatam*, a book that "tells the story of the Lord and His incarnations since the earliest records of Vedic history. It is verily the Krishna Bible of the Hindu universe."[6] In his commentary, he writes, "According to Vedic understanding, the entire universe is regarded as an ocean of space. In that ocean there are innumerable planets, and each planet is called a *dvipa*, or island. The various planets are divided into 14 *lokas*. As Priyavrata drove his chariot behind the sun, he created seven different types of

---

[4]Clarke, A. C. (1962). Hazards of prophecy: The failure of imagination. In *Profiles of the future* (p. 21). New York: Harper and Row.

[5]http://www.krishna.com/biography-srila-prabhupada

[6]Prabhu, A. A. Retrieved from http://www.srimadbhagavatam.org/introduction.html

oceans and planetary systems, known as Bhuloka."[7] Prabhupada then states that according to the Vedic tradition there are 400,000 species of living beings in the universe with humanlike forms, many of them advanced beyond us.

In 1970 Prabhupada wrote *Easy Journey to Other Planets*. In this book, he offers a description of the Hindu view of the physical universe. This description includes other planets and the expectation Hindus have for finding life on those planets. "In the material world the topmost planet is called Satyaloka, or Brahmaloka. Beings of the greatest talents live on this planet. The presiding deity of Brahmaloka is Brahma, the first created being of this material world. Brahma is a living being like so many of us, but he is the most talented personality in the material world. He is not so talented that he is in the category of God, but he is in the category of those living entities directly dominated by God…"[8] In this description of the material world, Brahma, the ruler of the planet Brahmaloka, is a created, living being, but not a God and not Brahman, the spirit and life force of the universe.

In keeping with the Hindu belief that the entire universe goes through cycles of creation and destruction and rebirth, Prabhupada identifies the lifespan for the universe as about 300,000 billion years: "The highest planet of the material universe, Brahmaloka, is also subjected to these modes of nature, although the duration of life on that planet, due to the predominance of the mode of sattva, is said to be $4,300,000 \times 1,000 \times 2 \times 30 \times 12 \times 100$ solar years… Lord Kr[i]s[h]na instructs that all the planets within the material universe are destroyed at the end of $4,300,000 \times 1,000 \times 2 \times 30 \times 12 \times 100$ solar years. And all the living beings inhabiting these material planets are destroyed materially along with the destruction of the material worlds …"

When all of these worlds are destroyed, the perfect yogis will leave their material bodies and will gain "entrance into the antimaterial universe or into the highest planets of the material sky. In the higher planets of the material world, the yogis can enjoy more comfortable and more pleasant lives for hundreds of thousands of years, but life in those higher planets is not eternal …" The yogis, in Hindu culture, are persons who actively practice one of the many forms of yoga, yoga being a discipline, or method of training, for transforming one's body and mind in order to unite with Brahman. This text refers to the most advanced yogis, the perfect yogis, who have achieved a high level of enlightenment and thus have earned long, comfortable lives on the highest of planets. Despite their perfection as yogis, even they, we learn, are nevertheless mortal.

The universe is presumed to be widely populated with living beings of all types. Any of those beings who have not yet escaped samsara may, upon reincarnation, occupy the physical body of a terrestrial life form and thereby return to Earth. Prabhupada writes, "Those who are not yogis but who die at an opportune moment due to pious acts of sacrifice, charity, penance, etc., can rise to the higher planets after death, but are subject to return to this planet [Earth]." But they do not have to return to Earth, because "There are not only infinite numbers of planets … but there are also infinite numbers of universes. All these infinite universes with their infinite planets within are floating on and are produced from the Brahman effulgence emanating from the transcendental body of Maha-Vishnu,

[7] http://www.hinduismtoday.com/modules/smartsection/item.php?itemid=3641
[8] From *Easy journey to other planets* by A. C. Bhaktivedanta Swami Prabhupada, courtesy of the Bhaktivedanta Book Trust International. Retrieved from www.Krishna.com

who is worshiped by Brahma, the presiding deity of the universe in which we are residing." As for getting to one of those other planets, that is possible if one's karma permits: "If one wants to go to the higher material planets, he can keep his finer dress of mind, intelligence and ego, but has to leave his gross dress (body) made of earth, water, fire, etc. When one goes to a transcendental planet, however, it is necessary to change both the finer and gross bodies, for one has to reach the spiritual sky completely in a spiritual form. This change of dress will take place automatically at the time of death if one so desires. But this desire is possible at death only if the desire is cultivated during life." The change of dress referred to here does not refer to clothing; rather, this change of dress means leaving behind one's mind, intelligence, ego, and physical body to become completely spirit.

Given that Hindus believe in an unfathomably large and eternal universe filled with life, an important question about Hindu beliefs needs to be addressed: Is the Earth special for Hindus? That is, can a living being achieve enlightenment and escape samsara if that being is incarnated anywhere other than the Earth? The answer to first question is 'no, the Earth is not special.' The answer to the second question is 'yes, escape from samsara can occur anywhere in the physical universe.' Hinduism does not demand that a life form be reincarnated on Earth in order to gain enlightenment. The Earth is important because Hindus live here, but the Earth is also one drop in the cosmic ocean, all of which is Brahman.

David Frawley writes, in "Hindu View of Nature" that for Hindus, "the Earth is sacred as the very manifestation of the Divine Mother. She is Bhumi Devi, the Earth Goddess. One of the reasons that Hindus honor cows is that the cow represents the energies and qualities of the Earth, selfless caring, sharing and the providing of nourishment to all. Hindu prayers are done at the rising of the Sun, at noon and at sunset, honoring the Divine light that comes to us through the Sun. Nature is always included in the Hindu approach. Even the great Hindu Yogis retire into nature to pursue their practices, taking refuge in the Himalayas and other mountains and wilderness areas where there is a more direct contact with the Divine."[9] A sacred Earth can, however, be a vital aspect of Hinduism but this idea is not identical with the Earth being the only place in the universe capable of being a sacred space. In principle, every planet on which Hindus live is sacred.

One of our basic questions about Hinduism has to do with whether living beings on other planets can be Hindus. The Hindu answer to this question would be 'yes.' Hinduism is not restricted to Earth and all beings throughout the universe could be practitioners of Hinduism. Similarly, Hindus would expect that Hinduism is a universal religion, that it is applicable throughout the universe and that any living being anywhere in the universe could choose to become a practicing Hindu.

Would Hindus need to proselytize, to convert extraterrestrials to Hinduism? No. According to an old Hindu saying, "There are multiple ways to the same place." These many ways include multiple Hindu paths as well as other religious approaches, so Hindus could choose to teach extraterrestrials one of the Paths to enlightenment, but they would not be compelled to do so. The Hindu monks of the Saiva Siddhanta Yoga Order, who are residents of the Hindu Monastery on the Hawaiian island of Kauai, have written, "Finally,

---

[9] Frawley, D. *Hindu view of nature* (Reprinted from Hindu Voice UK). Retrieved July 18, 2012, from http://www.vedanet.com/component/content/article/11-on-hinduism-sanatana-dharma/140-hindu-view-of-nature

it must be clearly understood that God and the Gods are not a psychological product of the Hindu religious mind. They are far older than the universe and are the fountainheads of its galactic energies, shining stars and sunlit planets. They are loving overseers and custodians of the cosmos, earth and mankind. The Hindu cosmological terrain envelops all of humanity. It is not exclusive. Hinduism has historically accepted converts from other religions and adoptives (those with no previous faith) into its knowledge and practices. ... A vedic rite ... declared, 'Why, born aliens have been converted in the past by crowds, and the process is still going on.' Each citizen of earth so interested has the option of entering the Hindu religion."[10] Presumably, spiritual citizens of other worlds would also be offered that option.

---

[10] *Gods and Gods of Hinduism*, the monks of the Saiva Siddhanta Yoga Order. Kauai's Hindu Monastery. Retrieved July 18, 2012, from http://www.himalayanacademy.com/resources/pamphlets/GodAndGodsOf Hinduism.html

17

Buddhism

> *In it there are thousands of suns, thousands of moons,*
> *thousands of inhabited worlds of varying sorts.*
>
> from *Anguttara Nikaya* (Numerical Discourses
> of the Buddha)

Buddhism emerged in India in the fifth century B. C. E. as an alternative to Hinduism for solving the central challenge of escape from samsara, the continuous cycle of rebirths. Historically, the Buddha (the "awakened one") was born in the sixth century B. C. E. as Siddhartha Gautama. He grew up as a prince in a royal family in what is now southern Nepal. After his personal search for a way to end the suffering of samsara, he discovered the knowledge that he was certain would liberate him from transmigration. Once enlightened, he taught five disciples about his insights in the form of "Four Noble Truths."

As understood by Buddhists, these truths, which are the essence of Buddhism,[1] were not uniquely revealed to the Buddha and thus are directly discoverable by all others who choose to follow the path of the Buddha. The Four Noble Truths are these: life is about suffering; suffering is caused by human desire; the cessation of desire—achieving nirvana—ends one's suffering; and the enlightenment necessary for reaching a state of nirvana is found by following the Noble Eightfold Path.

All of the many Buddhist scriptures are understood as the words and teachings of the one who became the Buddha or of other great Buddhist leaders. All of them were human and none of them are or became gods.[2] Some of these scriptures are quite old—the *Pali Canon*, which was first committed to writing in the first century B. C. E., is thought to preserve the teachings of the historical Buddha and his disciples as those words had been

---

[1] Gorski, E. F. (2008). *Theology of religions* (pp. 105–110). Mahwah, NJ: Paulist Press.

[2] http://buddhism.about.com/od/sacredbuddhisttexts/a/buddhist-scriptures.htm

D.A. Weintraub, *Religions and Extraterrestrial Life: How Will We Deal With It?*,
Springer Praxis Books, DOI 10.1007/978-3-319-05056-0_17,
© Springer International Publishing Switzerland 2014

passed down orally for the previous four centuries—but while they are sacred and revered, Buddhists do not consider them divine revelations.

Buddhism includes several cosmologies that may have once been accepted as precise descriptions of the actual physical universe. All of these are now best understood as metaphorical stages on which universal dramas play out. Even as metaphors, these cosmologies offer an understanding of Buddhism that informs us as to how Buddhists would react to announcements by scientists regarding the existence or absence of extraterrestrial life.

Buddhist cosmologies depict an incomprehensibly large universe populated throughout by sentient beings. The immeasurable vastness of space and immensity of time are filled with innumerable worlds. Among these worlds, humans are simultaneously both insignificant and all-important. With such a worldview, Buddhists take for granted the existence of advanced extraterrestrial life forms.

Buddhism has two major schools. Theravada Buddhism, which is dominant in southeast Asia and thus is known as the 'southern' school, advocates salvation through individual experience and enlightenment, i.e., enlightenment gained through one's own efforts. Mahayana Buddhism, which itself has fractured into numerous sects in 'northern' Asia, in regions such as China, Tibet, Mongolia, Korea and Japan, teaches salvation through meditation, chanting and prayer and focuses on the enlightenment of all beings rather than of the individual.

The two major schools of Buddhism put forward slightly different but mostly overlapping cosmologies, including the physical structure of the universe and the living beings that occupy the universe. According to the philosophies put forward by both schools, five different categories of sentient, living beings populate the universe. These five categories are souls damned to a next rebirth in a plane of torment and unhappiness, ghosts, animals, humans and deities. Buddhist deities, which are neither immortal nor spiritually enlightened, are sometimes further subdivided into the celestial *devas* and the lesser *asuras*. Including devas and asuras, these Buddhist categories include a total of six orders of living beings subject to karma and samsara.

Devas are non-human beings, invisible to humans who lack the power to see beings from other planes of existence. Devas possesses certain powers humans lack and enjoy more blissful lives than do humans. Some devas require food and drink; others do not. Devas enjoy happy lives of greater lengths than human lifespans, but devas are subject to samsara and thus will die and be reborn once their good karma has been exhausted. Many Buddhists believe that devas can grant favors to and protect highly religious humans. Asuras are also non-human beings who are more powerful than humans. They are described as impatient, hyper-competitive, paranoid, insincere, bellicose and prideful. Asuras live more pleasurable lives than humans but are envious of devas. Like devas, asuras are low-level gods subject to death and rebirth in other forms. In these important ways, Buddhist deities are distinct from the immortal *buddhas*, who have reached nirvana, and from the *bodhisattvas*, who have postponed final enlightenment in order to help those still caught in samsara. Though the deities may live for millions or billions of years, they nevertheless are subject to samsara.

Buddhists recognize three realms of existence, those of Desire, Form and Formlessness. Only the Realm of Desire, however, is located in the physical space of the universe that can be apprehended by human senses. The other two realms are accessible only through meditation.[3] Since the Realm of Desire is the only realm accessible to the unenlightened, it is the realm that can be understood as representing Buddhist ideas about the physical universe and about where extraterrestrials might be found.

The entire Realm of Desire is described as a thin, disk-shaped layer that lies atop a disk of water. The disk of water is much thicker than but identical in diameter to the Realm of Desire. The disk of water, in turn, lies atop a vast disk of wind. Thin is a relative term, however, as the "golden earth layer" atop which humans reside is described as 320,000 *yojanas* deep. A yojana is a measure of distance that was used in ancient India. One yojana might be described as the distance that a single bull could pull a cart or that an army could advance in 1 day. In modern terms, 1 yojana is approximately 10 km in length. Thus, the depth of the golden earth layer would be several million kilometers. Humans and animals inhabit this plane of existence, with the ghosts living 500 yojanas below the surface of the Earth and damned souls residing even lower. Mount Sumeru, which is located at the geographic center of the Realm of Desire, has roots that penetrate to a depth far below the surface of the golden earth layer and a summit that rises an equal distance above the surface. The Sun and the Moon circle Mount Sumeru around its elevation midpoint, supported, carried and contained by a great ring of wind.[4] The lowest-ranking devas, along with the Four Great Kings and 33 gods, reside atop the peak of Mount Sumeru. The asuras, in contrast, can be found almost anywhere, including at the base of the mountain, in the ocean or with the ghosts and damned souls far beneath the surface of the Earth. Various higher heavens, home to more advanced deities, are described as floating above Mount Sumeru in the upper regions of the Realm of Desire.

The Buddhist view of the physical world is well described in a widely and oft-quoted piece of text from the *Anguttara Nikaya* (Numerical Discourses of the Buddha), which describes the 'thousand-fold world' system. According to this text, the universe is full of inhabited worlds: "As far as these suns and moons revolve, shedding their light in space, so far extends the thousand-fold world universe. In it there are thousands of suns, thousands of moons, thousands of inhabited worlds of varying sorts … thousands of heavenly worlds of varying grades. This is the thousand-fold Minor World System. Thousands of times the size of the thousand-fold Minor World System is the twice-a-thousand Middling World System. Thousands of times the size of the Middling World System is the thrice-a-thousand Great Cosmos."[5]

Such a universe is unimaginably large. Even the observable universe accessible to modern astronomers, which is several tens of billions of light years in extent, would be a drop of water in the ocean that is the Buddhist's vision of the thrice-a-thousand Great Cosmos. The observable universe, however, is not identical with the entire universe. The most recent (2014) results from an international project known as BOSS (Baryonic Oscillation Spectroscopic Survey) suggest that the universe is infinite in spatial extent.

---

[3] Victor, A. G. *Basic Buddhism*. Retrieved from http://www.buddhismtoday.com/english/buddha/Teachings/basicteaching10.htm

[4] Sadakata, A. (1997). *Buddhist cosmology: Philosophy and origins* (pp. 25–41). Japan: Kosei Publishing.

[5] *Anguttara Nikaya* (vol. I, pp. 227, 228). As quoted in *Religious pluralism and the world community* (pp. 67–68). (1969). Netherlands: E. J. Brill Press.

The BOSS results would be consistent with Buddhist conceptions of this unimaginably large universe.

The Buddhist universe is also unimaginably ancient, even in comparison to the 13.7 billion year age assigned to the universe by modern astronomers, and time for Buddhists is cyclic. Like the living beings in the Buddhist universe, all the suns, moons and inhabited worlds in a thousand-fold world universe are believed to undergo a cycle of creation and destruction (reincarnation on a cosmic scale). The length of time for one of these cycles is a *kalpa*. Eighty successive kalpas form one *mahakalpa*, an enormous period of time often defined only through comparisons, such as the length of time required to erode a rock one cubic yojana in size by wiping it with a piece of cotton once every 100 years.

Many attempts have been made in Buddhist literature to classify the vast array of beings on these many inhabited worlds. In a 1958 report to UNESCO entitled "Buddhism and the Race Question," G. P. Malalasekera, who was then Dean of the Faculty of Oriental Studies at the University of Ceylon (known after 1972 as the University of Sri Lanka), and K. N. Jayatilleke, then a Lecturer in Philosophy at the same institution, write about how humans fit into these classification schemes. "One such classification speaks of human beings, as well as some of the higher and lower beings, as falling into the class of beings who are different and distinguishable from each other in mind and body. There are other classes where the beings are different in body but one in mind. Yet others are alike in body but different in mind, while there are some who are alike both in body and in mind. A further set of four classes of beings are mentioned who are formless. All these are described as the several stations which the human consciousness can attain and find renewed existence after death. ... The human worlds are always represented as standing midway in the hierarchy of worlds. Life in these human worlds is a mixture of the pleasant and the unpleasant, the good and the evil, while the pleasant and good traits are intensified in the higher worlds and the unpleasant and evil in the lower."[6]

Whatever these beings may be like, Nyanaponika Thera, a Sri Lanka-ordained monk and co-founder of the Buddhist Publication Society, writes that in the most important way, they are just like us, as they "are subject to the law of impermanence and change." In unimportant ways, of course, they may appear different from us, Thera writes, but other worlds may also be very similar to ours. "The inhabitants of such worlds may well be, in different degrees, more powerful than human beings, happier and longer-lived. Whether we call those superior beings gods, deities, devas or angels is of little importance, since it is improbable that they call themselves by any of those names. They are inhabitants of this universe, fellow-wanderers in this round of existence; and though more powerful, they need not be wiser than man. Further, it need not be denied that such worlds and such beings may have their lord and ruler. In all probability they do. But like any human ruler, a divine ruler too might be inclined to misjudge his own status and power, until a greater one comes along and points out to him his error, as our texts report of the Buddha."[7]

Perhaps surprisingly, although Buddhist cosmologies imply that every world system is inhabited, buddhas themselves are thought to be rare, as measured across both space and time.

---

[6] Malalasekera, G. P., & Jayatilleke, K. N. (1958). *Buddhism and the race question* (p. 33). UNESCO.
[7] Thera, N. (2008). *Buddhism and the God-idea* (online edition, p. 3). Buddhist Publication Society. Retrieved from http://www.scribd.com/doc/37750186/Buddhism-and-the-God-Idea-Selected-Texts

They are more rare, however, in Theravada than in Mahayana cosmology. In Theravada cosmology, buddhas are rare in any sizeable volume of the universe as well as across any large expanse of time. A buddha can be expected to appear just once during a single mahakalpa and in only one of the billion world systems of the thousand-fold world universe, as described in Theravada cosmology. Some more optimistic accounts describe as many as 1,000 buddhas appearing in each mahakalpa, but even then the next buddha, Maitreya, is not expected to appear for another 5.67 billion years.[8] Whether one takes an optimistic or pessimistic view of the prevalence of buddhas, the incredible unlikelihood of any single world ever hosting a buddha would seem to place the Earth in an extremely privileged position, a notion with important consequences for the encounter between Theravada Buddhist beliefs and any future discoveries about extraterrestrial life.

The Mahayana conception of the universe is far vaster than that of Theravada, emphasizing the universality of Buddha. Earth occupies a far less privileged position in Mahayana than in Theravada cosmology. Although Mahayana scriptures typically affirm the general traits of the Mount Sumeru-centered world system, they rework them into a variety of much vaster cosmologies, many of them centered on the metaphor of a lotus flower. In one, the universe is conceived of as a thousand-petaled lotus flower with ten billion world systems on every petal, for a total of ten trillion worlds. Another describes a cosmic lotus with seed receptacles as numerous as "atoms in indescribably many buddha-fields," each containing a "fragrant ocean" itself replete with world systems. In both, all the worlds are conceived of as creations of *Vairocana* ("Illuminating All Places"), the transcendent personification of universal buddha-nature (*dharmakaya*). That is, this Vairocana represents the cosmic possibility of Buddhahood, a potentiality that would exist even had the historical Buddha never lived. The omnipresent Vairocana incarnates into countless buddhas and bodhisattvas, one for every world, presenting a far more optimistic view of existence, in which all of the life forms in the universe have access to nirvana by way of their own particular manifestations of Vairocana.[9]

Both the Theravada and Mahayana cosmologies assume that extraterrestrial life forms exist, but for practical purposes any such aliens are mostly irrelevant to human existence, although stories do exist of buddhas and lesser beings from other world systems occasionally visiting Earth, for instance to pay homage to our own world's buddha.[10] The existence of extraterrestrial life, however, is one of the key doctrines of the Pure Land sects of Mahayana Buddhism. Pure Land Buddhism revolves around the prehistoric buddha Amitabha, who, while on the path towards enlightenment, made a vow that he would liberate from suffering any being who called on his name. After attaining buddhahood, Amitabha is said to have fulfilled his vow by creating the "pure land" of *Sukhavati*, a world devoid of suffering and maximally tailored to spiritual pursuits, in which his followers

---

[8] Sadakata, A. (1997). *Buddhist cosmology: Philosophy and origins* (pp. 96–97, 108–109). Japan: Kosei Publishing.

[9] Ibid (pp. 144–154).

[10] De Bary, Wm. T., & Bloom, I. (1999). *Sources of Chinese tradition Volume One* New York: Columbia University Press.

may pray for rebirth to achieve rapid enlightenment under his tutelage.[11] Despite its paradisical attributes, Sukhavati is described as a place within our universe located "ten myriads [100,000] of a hundred millions of buddha-lands to the west."[12] Although Pure Land Buddhism focuses mainly on Amitabha Buddha and the land of Sukhavati, other pure lands with their own buddhas are described as well; for instance, the relatively nearby Realm of Profound Joy is described as "one thousand buddha-lands to the east."[13]

Today, Buddhists almost universally acknowledge the legitimacy of modern science, though two distinct approaches exist in harmonizing modern science with Buddhist doctrine. The more conservative viewpoint holds that traditional Buddhist cosmology has always reflected the reality of the universe, though in an approximate rather than a literal manner. In this view, the most important element of any of the Buddhist cosmologies is the vision of a vast, complex and extremely ancient universe, a vision Buddhists would argue has been confirmed by modern science, even if at the expense of specific details. The more radical view makes no attempt to reconcile Buddhism and science, but rather separates them into distinct areas of inquiry. This approach dates back to the second-century Buddhist philosopher Nagarjuna, who argued for a "two-truth" epistemology. One kind of truth is pragmatic, derived from human thought and reason, including all human concepts and all human language and including all aspects of Buddhism. Such truths are considered secondary, or conventional, truths. In contrast, absolute truths exist independent of human attempts to describe them.[14] In this spirit, the fourteenth (and current) Dalai Lama Tenzin Gyatso has stated that the Buddha, in promoting the Mount Sumeru-centered physical cosmology, was presenting a conventional truth, one that is dependent on human concepts and language, and that Buddhists are not required to maintain a belief in this cosmology. The possible error, with respect to the discoveries of modern science, in the cosmological doctrine found acceptable two-and-a-half thousand years ago by the Buddha does not undermine the central message of Buddhism, however, for "the purpose of the Buddha's coming to this world was not to measure the circumference of the world and the distance between the earth and the moon, but rather to teach the Dharma, to liberate sentient beings, to relieve sentient beings of their sufferings."[15]

Due to the overwhelming acceptance of scientific knowledge by most modern Buddhists, the discovery of extraterrestrial life would not in itself pose any difficulties for Buddhism. Although a literal reading of traditional cosmology does conflict with many specific astronomical findings, particularly in its vision of a geocentric world system and the numbers of and distances to other world systems, these conflicts have already been

---

[11] Jones, C. B. (2003). Transitions in the practice and defense of Chinese pure land Buddhism. In S. Heine, & C. S. Prebish (Eds.), *Buddhism in the modern world: Adaptations of an ancient tradition*. United States: Oxford University Press.

[12] Sadakata, A. (1997). *Buddhist cosmology: Philosophy and origins* (p. 115). Japan: Kosei Publishing.

[13] Ibid.

[14] Ingram, p. O. (2008). *Buddhist-Christian dialogue in an age of science* (p. 48). United States: Rowman and Littlefield Publishers.

[15] Lopez, D. S. (2008). *Buddhism and science: A guide for the perplexed* (p. 63). Chicago: University of Chicago Press.

resolved, and the existence of extraterrestrials would not in itself produce any new difficulties. Those espousing a more conservative view would probably take such a discovery as confirmation of the ultimate compatibility between traditional religious and scientific models of the universe.

Whether the life we discover includes sentient beings probably would not change the outcome, since traditional Buddhist cosmology posits a multitude of world systems inhabited by the same five orders of life as our own. For those more inclined towards the "two-truth" approach, the discovery of extraterrestrial life might generate some enthusiasm, but almost certainly would not be considered directly relevant to the larger message of Buddhism. Either way, such findings would easily be accommodated into Buddhist religious beliefs, as further indicated by the findings of the Peters ETI Religious Crisis Survey Report.[16] Peters reports that out of eight religious categories, Buddhists most strongly *disagreed* with the statement "Official confirmation of the discovery of a civilization of intelligent beings living on another planet would so undercut my beliefs that my beliefs would face a crisis."

Furthermore, beyond mere acceptance, certain forms of Buddhism—specifically, Pure Land Buddhism—might actually thrive as a result of such a discovery. Barring incredible technological innovations permitting rapid interstellar travel, direct physical contact of humans with extraterrestrial life forms is unlikely within the foreseeable future. By contrast, Pure Land Buddhism offers an opportunity for direct communication with an alien intelligence (if Amitabha Buddha is reinterpreted as such) by way of meditation and the subsequent rebirth of one's spiritual essence in a world that might require countless human lifetimes to reach by technological means. No Buddhist soul would be able to interact with an extraterrestrial being while incarnated in a human body. If that soul, however, passed from the body of a human on Earth to the body of another being in one of the Pure Lands, that particular soul would achieve physical contact with an otherworldly being. In the wake of the discovery of extraterrestrial life, Pure Land Buddhism could easily ride the wave of public excitement and curiosity by offering followers the possibility of a far more intimate encounter with alien life forms, albeit only in one's next lifetime, than science could realistically hope to provide.

The possibility that decades of thorough scanning of hundreds of millions of exoplanets might fail to uncover any evidence of extraterrestrial life would be fine for some Buddhists but could be difficult for others to handle. While the evidence for the absence of extraterrestrials in astronomy surveys, no matter how extensive, would not disprove the existence of life beyond the range that telescopes can observe or of life-forms too unlike terrestrial organisms to produce recognizable signs, they would strongly suggest that if extraterrestrial life exists at all it is extremely rare. For followers of the "two-truths" approach, for whom Mount Sumeru-centered cosmology is at best a beautiful fiction, this would not pose a particular problem; however, for those who affirm the continued relevance of traditional cosmological views, such findings could be problematic.

---

[16] Peters, T., & Froehlig, J. *The Peters ETI religious crisis survey* (p. 7). Counterbalance Foundation. Retrieved from http://www.counterbalance.org/etsurv/index-frame.html

According to the *Abhidharmakosa*, the wind circle that forms the foundation of the Mount Sumeru realm is held together by the collective karma of living beings.[17] Furthermore, the cosmic process of creation and destruction itself is driven by this collective karma. The dissolution of the universe begins when all damned souls expend their negative karma. This event causes the various lower planes of existence in which they live to vanish. This process repeats itself for every level of the Realm of Desire until the entire physical world is destroyed. During this cosmic collapse, beings in the higher planes of existence will still generate low levels of karma worthy of lower rebirth in realms that no longer exist. This karma is therefore stored as potential energy in the fabric of reality and its effects postponed until, after a period of 20 kalpas, it again manifests as a tiny wind, from which the wind and water circles the golden earth layer. Eventually the entire Mount Sumeru world would gradually return to being populated by those beings that had not yet expended their karma in the previous incarnation of the physical universe.[18]

From the perspective of this traditional cosmology, the very possibility of the cyclic birth and rebirth of the physical universe depends on the presence within the universe of living beings. Without living beings, the unexpended karma needed to trigger the continual rebirth of the physical universe would not be created and so the universe could not return to existence. Those who still place some weight in such doctrines as an approximately accurate model of the universe might find strange the idea that the collective karma of the inhabitants of just one planet could generate so vast a universe; on the other hand, this might simply be taken to illustrate the incredible power of karma. Alternatively, the absence of life elsewhere in the universe could also be harmonized with Buddhism by positing that we live in an early stage of the current universe in which not all beings left over from the last cycle have yet been reborn, so that the entire cosmos will eventually become populated even if it currently is not. This argument would more fully reconcile the observed absence of extraterrestrial life with scriptural assertions as to the multiplicity of inhabited world systems. It does, however, have the disadvantage of conflicting with traditional Buddhist models of time, which typically place the current age in a late stage of the existence of this universe, a stage marked by decreasing lifespan and the decay of Buddhist teachings.[19]

If scientific findings that reveal no signs of extraterrestrial life were taken as an absolute indication that no form of life currently exists anywhere else in the universe, Buddhists would either have to significantly amend the features of their biocentric cosmology or dispense with that cosmology entirely as a model of objective reality, as already done by some prominent leaders such as the Dalai Lama. This conclusion, of course, would only be based on a study by astronomers of a relatively small sample of the observable universe, and a conclusion that such a study provided absolute evidence that no extraterrestrial life exists in the entire universe would be scientifically absurd. Therefore, even if science eventually demonstrates that life as we know it is extremely rare elsewhere in our part of the universe, scientists will never be able to state that life beyond Earth, in any form, absolutely

---

[17] Sadakata, A. (1997). *Buddhist cosmology: Philosophy and origins* (p. 69). Japan: Kosei Publishing.

[18] Ibid (pp. 102–104).

[19] Ibid (pp. 105–107).

does not exist. The conclusion that life is rare would appear to leave enough uncertainty for Buddhists to continue to believe in extraterrestrial life. With regard to Pure Land Buddhism, the fact that Amitabha's world of Sukhavati is traditionally located in an extremely distant part of the universe makes its existence essentially unfalsifiable and therefore a matter of faith rather than of scientific inquiry. Therefore, although the apparent scarcity of life in the universe would certainly prove surprising to many Buddhists, even the more conservative strands of Buddhist thought could probably adjust to such findings with only minimal refinements to their views.

18

Jainism

> *The number of suns in the entire region inhabited by humans*
> *is 2 + 4 + 12 + 42 + 72 = 132.*
>
> Sutra 4.14 of the *Tattvārtha Sutra*

Jainism arose in the sixth century B. C. E., at about the same time and in the same place and for the same reasons as Buddhism, though "according to its own traditions, the teachings of Jainism are eternal, and hence have no founder."[1] The sage Mahāvīra (meaning "great hero" in Sanskrit; either 599–527 B. C. E.[2] or 563–483 B. C. E.[3]), a contemporary of the Buddha, was the last of the "spiritual victors" known as Jina, the 24 enlightened humans who, during the most recent descending cycle of time, followed and taught others how to follow the path across *samsara* (the river of suffering) toward *moksha* (liberation from the cycle of rebirth), thereby reaching omniscience by freeing their souls from all karma. Mahāvīra's disciples, the followers of the Jina and the eternal concepts of Jainism, are known as Jains. Today, most of the approximately four million Jains worldwide live in India.

The fundamental principle underlying Jainism is non-violence toward all living beings, whether ants, elephants, humans, plants or bacteria, as Jains believe everything that lives has a soul that is eternal and potentially divine. Even water and rocks are treated with extreme care, as non-living objects in every part of the universe can be homes to countless, microscopic living things. As a result, Jains are strict vegetarians, eating only the byproducts (e.g., nuts, fruits, vegetables, milk) of living beings and avoiding all acts of violence.

---

[1] Webb, M. O. (2005). *Internet Encylopedia of Philosophy*. Retrieved from http://www.iep.utm.edu/jain/
[2] Umāsvāti (1994) (N. Tatia, Trans.) *Tattvārtha Sutra (That Which Is)* San Francisco: HarperCollins.
[3] Kelly, J. *Jainpedia: the Jain universe online*. Retrieved from http://beta.jainpedia.org/themes/people/jina/mahavira.html

To modern Jains, Jainism is more than a religion. According to K. V. Mardia, writing in *The Scientific Foundations of Jainism*, Jainism "tries to give a unified scientific basis for the whole cosmos." Mardia notes that Jain science is qualitative and Jain ideas and modern science are almost always in agreement. "Jainism is science with religion. Every aspect of Jainism is based on understanding the cosmos, and the living and non-living entities in it."[4] For example, from a Jainist perspective modern science has revealed things that Jains claim to have known for thousands of years, for example the fact that microbes exist in drops of water, lumps of dirt and air.

If the Jaina concepts of cosmology and of the many types of living beings are understood at least in part as metaphorical, then the concept of extraterrestrial life has a basis in Jainism. In Jainist thought, both the universe and the natural laws of the universe are eternal and uncreated and therefore have no creator. Those laws, rather than a deity, control the structure and nature of the universe. For Jains, space is partitioned into two regions: cosmic space and transcosmic space. Cosmic space contains the principles of motion and rest and thus is inhabitable by matter, souls and living beings; by analogy, we can equate cosmic space to our physical universe. Transcosmic space completely surrounds cosmic space. It is a boundless "'space without worlds,' which is devoid of souls, matter, and time"[5] and extends infinitely far in all directions.

The inhabitable part of space, cosmic space, has four realms consisting of three principal realms plus a fourth realm that sits like a small cap atop the highest of the three principal realms. The lowest of the three principal realms is the region of hells; the middle principal realm is the region inhabited by humans; the upper principal realm is the region of heavens. Each region has a very explicitly defined size and shape. The lower region can be pictured as an inverted, tapered, four-sided vase. The middle region can be visualized as a smaller, four-sided, tapered vase, resting right-side up on the narrow, upper (inverted) end of the lower region. That is, the narrow end of the lower realm points up while the narrow end of the middle realm points down, with the two narrow ends touching. Finally, the upper region, which is the same size as the middle region, can be imagined as an inverted, tapered, four-sided vase that rests on the top of the middle region, with their respective four-sided 'tops' in contact. Sitting just above the highest realm is the Realm of the Jinas. Those souls that have been liberated from samsara dwell in the Realm of the Jinas.[6]

The lower world is a seven-layered underworld in which the wicked are punished and purified before they are reborn. Humans live in the middle, or earthly, realm. The Jain conception of the middle realm is one of 90 circular, concentric oceans and continents. The central continent (*Jambū-dvīpa*), the second continent (*Dhātakīkhanda*), and the inner half of the third continent (*Puṣkara-dvīpa*) are the Two and A Half Continents where

[4] Mardia, K. V. (1996). *The scientific foundations of Jainism* (pp. 101, 111). Delhi: Motilal Banarsidass Publishers.

[5] Jaini P. S. (1983). *Karma and rebirth in classical Indian traditions*. In W. Doniger (Ed.) Delhi: Motilal Banarsidass Publishers. Appendix 1. Retrieved from http://personal.carthage.edu/jlochtefeld/picturepages/smallstuff/jaincosmology.html

[6] Gorski, E. F. (2008). *Theology of religions* (pp. 122–129). New York: Paulist Press.

humans can live.[7] The third and highest of the three parts of the physical world is heaven, where only gods may reside (though not all gods reside in the upper world). The upper realm consists of as many as 39 separate heavenly abodes.

Jains classify living beings in the earthly realm according to the number of senses these creatures possess for interacting with the physical universe. Plants have only a single sense, touch. Worms have two senses, touch and taste. Termites have three senses, as they possess the ability to smell as well as the abilities to touch and taste. Butterflies are an example of a species with four senses, as they have the sense of sight, in addition to the three senses of touch, taste and smell. Humans and other mammals have five senses: touch, taste, smell, sight and sound. In the ladder of living things for Jains, a greater number of senses corresponds to a higher level of consciousness and also corresponds, in Jaina thought, to a more advanced evolutionary status. The concept of evolution is consistent, qualitatively, with Jainism, although Jainism adds an evolutionary element of critical importance to Jain theology. Only living beings that, through the course of evolution, have been born on the step on the evolutionary ladder at which they possess all five senses are capable of conscious thought, and only souls capable of conscious thought are able to follow the path toward enlightenment.

For Jains, all differences between individual living beings, whether these differences can be attributed to wealth, health, size, color, shape or power, are due to differences in the karma possessed by those beings. Karma is the original cause of everything, and all phenomena are due to the effects of good and bad karma. Even gods, who exist but do not have control over the entire universe, are subject to karma and rebirth. Each living soul, affected by karma from previous lives, remains attached to the physical world. In fact, for Jains, karma is a material substance that attaches and reattaches a soul to a physical body. Therefore, to free oneself from transmigration and achieve the condition of perfection, thereby becoming a liberated soul (a *siddha*), a Jain must free the soul from all karma, good and bad. Liberation is achieved through self-control, non-violence, truthfulness and detachment from worldly possessions.

According to Jainism, if a soul is not fully unburdened of all forms of karma at the time of the death of the living being in which it had existed, it is reborn. At each incarnation, a soul is reborn as one of four kinds of beings (heavenly, human, infernal, animal or plant) in one of the three principal realms of the universe; however, only humans—not even gods—can achieve omniscience and become siddhas; thus, being born as a human in the middle realm is critical for becoming a siddha and potentially reaching nirvana.

Each soul has an infinite amount of time to become a siddha, since the Jain universe is eternal and time is cyclical. Each cycle of time includes an enormously long period of decline followed by an equally long period of ascent. During periods of decline, people are corrupt and immoral and religion is necessarily introduced to provide foundational ethics for individuals and society. During periods of ascent, people are happy, wise, virtuous and without need of religion. As the cosmic wheel of time turns, each cycle transitions smoothly into the next. As Jains believe that Jainism was recently (re)introduced into the

---

[7] Balibar, N. *Jainpedia: the Jain universe online*. Retrieved from http://beta.jainpedia.org/themes/principles/jain-universe/the-three-worlds/contentpage/3.html

universe by the 24 Jina, the current epoch is necessarily one of spiritual decline and hence one in which religion is present.

A soul that is transmigrating from one life in one living being to its next life in another living being may transit a distance across the physical cosmos that is limited only by the size of the cosmos itself. While transiting, a soul must travel in straight lines, either east-west so that the soul remains at the same level in the same realm of the universe, or north-south, traveling upwards or downwards through one or more layers of the lower, middle and upper realms. One complete transmigration event can encompass no more than three turns and four straight-line motions through the cosmos, but can transport a soul from one end of the Jaina universe to the other. The transmigration of a soul happens almost instantaneously, no matter how great the distance it must travel. The duration of a transit is one time unit for each straight line movement, and therefore up to four time units in total. One time unit is described as "too small to be measureable in the ordinary sense."[8] An innumerable number of time units make up the smallest unit of countable time, which is 45/262,144 of a second. If cosmic space is analogous to the physical universe, transmigration allows Jaina souls to relocate across tens of billions of light years in a fraction of a second; this property of super-light speed movement would allow living beings to inhabit every corner of the universe.

According to Jainist cosmological beliefs, the physical structure of the universe—the size and the shape of cosmic space—clearly is not identical with the modern scientific understanding of the structure of the physical universe. Most Jains, however, continue "to learn and teach Jain cosmology for its own sake without rejecting modern science."[9] This approach allows Jains to take the position that their own cosmology is somehow compatible with that of modern science. In Sutra 4.14 of the *Tattvārtha Sutra*, we learn that "There are two suns above Jambū Island, four over Lavana Ocean and twelve over Dhātakīkhanda Island. ... The number of suns in the entire region inhabited by humans is $2 + 4 + 12 + 42 + 72 = 132$. The number of moons in the region inhabited by humans is the same as that of suns. ... Each moon has an entourage of 28 constellations, 88 planets and $66,975 \times 10^{14}$ stars."[10] We can easily conceptualize these assertions as compatible with a universe teeming with galaxies, each one of which contains billions of stars.

The obvious way to achieve that compatibility would be to identify the middle realm as a metaphor for all places in the known universe compatible with the existence of any kind of living being. Thus, the middle realm could represent or encompass all inhabitable planets or planet-like locations in the universe. When viewed through a metaphorical lens, nothing about Jainist concepts confines life to the single planet Earth, as souls could bond with any non-living matter and the matter containing souls could evolve in a biological sense anywhere in the physical universe. Similarly, transmigration of a soul could move

---

[8] Umāsvāti 4.15 (p. 104).

[9] Balibar, N. *Jainpedia: the Jain universe online*. Retrieved from http://beta.jainpedia.org/themes/principles/jain-universe/contentpage/4.html

[10] Umāsvāti 4.14 (p. 103).

that soul to a new physical body located anywhere in the universe. When understood this way, the discovery of life beyond the Earth would not trigger any problems for Jains, and Jainism could be understood as a universal path for all living beings in the universe to escape samsara.

The single, problematic issue for Jains might be the belief that one's life as a human is the only opportunity, during the endless cycle of reincarnations, when a soul can learn how to live out a life of self-control, non-violence, truthfulness and worldly detachment that leads to shedding the burden of all karma, to omniscience and to an escape from samsara. Either life on Earth as a human is truly a unique opportunity for a Jain, and thus the Earth is a place of unparalleled importance in the universe, or the idea of what it means to be human would need to be broadened to include all extraterrestrial life forms that are considered higher than the other mammals and all other animals but lower than the gods.

19

Sikhism

So many worlds beyond this world—so very many!

from the *Sri Guru Granth Sahib*

Sikhism is a monotheistic religion that emerged at the close of the fifteenth century in what is now the Punjab province of northern India. Today, Sikhs make up just over 60 % of the population of the Indian Punjab and number about 27 million adherents worldwide. Like Hindus, Buddhists and Jains, Sikhs believe in samsara, the cycle of birth and rebirth. According to Sikh belief, when humans break out of this cycle their souls become one with God. Sikhs can reach this moment of spiritual unification by demonstrating their devotion to and knowledge of God and by being charitable and performing acts of service to their communities.

According to Sikh beliefs, the first living Guru, Guru Nanak (b. 1469 C. E.), brought and delivered the Word of God to Earth beginning in 1499 C. E. In 1708 C. E., the tenth and last Guru in human form, Guru Gobind Singh, left behind the holy text known as *Sri Guru Granth Sahib*. These ten human prophets and spiritual teachers established the Sikh religion. This Granth (holy book), which contains the words spoken by the ten human Gurus, is treated by Sikhs as the eleventh, and last, Guru. In this way, the Word of God first brought to Earth in spoken form by the first Guru and then spoken by him and his nine human Guru successors for two centuries was enshrined for all future times in the form of the written Word. The Sri Guru Granth Sahib is now considered the "Supreme Spiritual Authority and Head of the Sikh religion."[1]

Sikhism is a very human-focused religion that emphasizes how humans should live and interact. As such, Sikhism includes five articles of faith that all are extremely human-centric. They are often identified as the five K's of Sikhism. One of the articles of faith

---

[1] http://www.sikhs.org/granth.htm

D.A. Weintraub, *Religions and Extraterrestrial Life: How Will We Deal With It?*,
Springer Praxis Books, DOI 10.1007/978-3-319-05056-0_19,
© Springer International Publishing Switzerland 2014

involves what a Sikh may not do: a devout Sikh may not cut his or her hair (uncut hair is referred to as *kesh*), including all head and body hair. Male Sikhs will wear a turban (the requirement to wear a turban is optional for Sikh women), as an outward demonstration of their faith. The other four articles of faith identify objects a devout Sikh must wear or carry: a small wooden comb (*kanga*) worn in their hair, an iron or steel bangle worn on the right arm (*kara*), a short, ceremonial dagger (*kirpan*) and a loose undergarment (*kachhara*).

Sikhism attempts to regulate human behavior by rejecting any and all differences of caste, creed, race or sex; by promoting the full equality of women; by urging that followers overcome the vices of lust, anger, greed, pride and worldly attachment; by encouraging believers to avoid alcohol, meat, tobacco and drugs; and by developing and nurturing the positive traits of truth, compassion, contentment, humility and love. Sikhism promotes daily meditation and scriptural study and rejects practices identified by Sikhs as superstitions and rituals, such as fasting, animal sacrifices, pilgrimages and worshiping the dead or idols.

In Sikh cosmology, the one God who rules over all of creation created the entire universe, including myriad planets and solar systems. In the words of the Sri Guru Granth Sahib,

> For endless eons, there was only utter darkness.
> There was no Earth or sky; there was only the infinite Command of His Hukam [*divine command or will*].
> There was no day or night, no Moon or Sun; God sat in primal, profound Samaadhi [*meditative position*].
> There were no sources of creation or powers of speech, no air or water.
> There was no creation or destruction, no coming or going.
> There were no continents, nether regions, seven seas, rivers or flowing water.
> There were no heavenly realms, earth or nether regions of the underworld.
> There was no heaven or hell, no death or time.
> There was no hell or heaven, no birth or death, no coming or going in reincarnation.
> There was no Brahma, Vishnu or Shiva. ...
> When He so willed, He created the world.
> Without any supporting power, He sustained the Universe. ...
> He created the creation, and watches over it;
> the Hukam of His Command is over all.
> He formed the planets, solar systems, nether regions and brought what was hidden to manifestation.[2]

Closer to home, God filled in the details, including creating life:

> He established the earth, the sky and the air, the water of the oceans, fire and food.
> He created the moon, the stars and the sun, night and day and mountains;
> He blessed the trees with flowers and fruits.
> He created the gods, human beings and the seven seas;
> He established the three worlds [*life in water, on land and in the air*].[3]

---

[2] *Sri Guru Granth Sahib* (pp. 1035–1036), from the Khalso Consensus (Singh Sahib Dr. Sant Singh Khalsa, Trans.). http://www.srigranth.org/servlet/gurbani.gurbani
[3] Ibid (p. 1399).

While this passage could be understood as describing the millions of forms of life in the oceans, on the land, and in the skies of just the Earth, in the context of other passages, it can and should be understood as a description of life created throughout the universe. For example, in the following passage, the Sri Guru Granth Sahib teaches about the millions of created worlds populated by "beings of so many descriptions":

Many millions are the moons, suns and stars.
Many millions are the demi-gods, demons and Indras ...
Many millions are the beings of so many descriptions.
Many millions are made long-lived. ...
Many millions are the fields of creation and galaxies.
Many millions are the etheric skies and the solar systems.
Many millions are the divine incarnations.
In so many ways, He has unfolded Himself.
So many times, He has expanded His expansion.
Forever and ever, He is the One, the One Universal Creator.
Many millions are created in various forms.[4]

Taken together, these scriptural verses make clear that, as understood by Sikhs, the God worshiped by Sikhs created the entire universe and all living things throughout the universe. More explicitly, these passages explain that God created living beings throughout the universe and not just on Earth. The Sri Guru Granth Sahib re-emphasizes this concept by instructing Sikhs that living beings exist in worlds far from the Earth:

I see none as great as You, O Great Giver;
You give in charity to the beings of all the continents, worlds, solar systems, nether regions and universes.[5]

This idea is repeated and reinforced in another verse:

Creating the world of the nine regions, O Lord,
You have embellished it with beauty.
Creating the beings of various kinds,
You infused Your power into them.[6]

And the poetry of the Sri Guru Granth Sahib emphasizes that through life in all places, the Lord is in all places:

Many solar systems, many galaxies.
Many forms, colors and celestial realms.
Many gardens, many fruits and roots.
He Himself is mind, and He Himself is matter.
Many ages, days and nights.

---

[4] *Sri Guru Granth Sahib* (pp. 275–276).
[5] Ibid (p. 549).
[6] Ibid (p. 1094).

Many apocalypses, many creations.
Many beings are in His home.
The Lord is perfectly pervading all places.[7]

The Sri Guru Granth Sahib continually and repeatedly affirms that all creatures every-where were created by the one God:

So many worlds beyond this world—so very many!
What power holds them, and supports their weight?
The names and the colors of the assorted species of beings
Were all inscribed by the Ever-flowing Pen of God.[8]

You alone are the Lord of so many thousands of worlds.[9]

In the realm of karma, the Word is Power.
No one else dwells there,
Except the warriors of great power, the spiritual heroes. ...
The devotees of many worlds dwell there. ...
There are planets, solar systems and galaxies.
If one speaks of them, there is no limit, no end.
There are worlds upon worlds of His creation.
As He commands, so they exist.
He watches over all, and contemplating the creation, He rejoices.[10]

In the heavenly paradise, in the nether regions of the underworld,
on the planet Earth and throughout the galaxies,
the One Lord is pervading everywhere.[11]

Despite the repeated emphasis by the Sri Guru Granth Sahib on the abundance of life throughout the created universe, Sikhism also has a geocentric bias. For example, not only is the Earth the location where human beings live and their deeds and actions judged, the natural laws and order of the universe that lead toward nirvana (*Dharma*) may have been established foremost on Earth:

In the midst of these, He established the Earth as a home for Dharma.
Upon it, He placed the various species of beings.
Their names are uncounted and endless.
By their deeds and their actions, they shall be judged.[12]

---

[7] Ibid (p. 1236).

[8] Ibid (p. 3).

[9] Ibid (p. 727).

[10] Ibid (p. 8).

[11] Ibid (p. 207).

[12] Ibid (p. 7).

Do these words in the Granth indicate that Earth is the only home for Sikh Dharma? Could the actions of a Sikh be judged if that person lived in a distant part of the universe rather than on Earth? Answers to these questions might be found in thinking about the five articles of faith for Sikhs.

The philosophy and practice of Sikhism would have no meaning for an intelligent being who is hairless and has no appendages on which to place a bracelet, wear the two-legged undergarments, or strap on a sacred sword, i.e., for any creatures other than humans or a human-like species. While Sikhism could easily be practiced by humans or human-like creatures anywhere in the physical universe, the five articles of faith for Sikhs appear to establish a set of religious practices that are exclusively for living beings who are physically and culturally similar to us. With this understanding, the Earth could in fact be the exclusive location in the universe for Sikh Dharma.

Sikhism appears to be a terrestrial religion that offers a path that humans can follow in order to find God and escape samsara. Sikhism, however, is not a universal religion, even if the God of Sikhism, as understood by Sikhs is a universal God, "the Master of all worlds ... the Creator of all worlds."[13]

Sikhs clearly assume that other living beings exist in abundance in the universe. These beings include intelligent creatures capable of following their own paths of revealed wisdom, paths that might be different from the Sikh Dharma established on Earth, that will unite their souls with God. The discovery of extraterrestrial life would not be a surprise to Sikhs and would serve to reinforce their beliefs. The failure of humans to discover definitive evidence of extraterrestrial life, however, would not put a damper on the beliefs of the faithful, as the universe is so large that God's millions of worlds could nevertheless be separated by distances so large as to place our closest neighbors much too far away to be discoverable by us in the foreseeable future.

[13] Ibid (pp. 1079, 1095).

20

Bahá'í Faith

> *Know thou that every fixed star hath its own planets,*
> *and every planet its own creatures, whose number*
> *no man can compute.*
>
> Bahá'u'lláh, in *Gleanings from the Writings of Bahá'u'lláh*

The Bahá'í Faith, which today has five to six million followers worldwide, emerged in the mid-nineteenth century when a Persian nobleman, known as Bahá'u'lláh (1817–1892 C. E.), had a vision of God's will for humanity and proclaimed himself one of God's Messengers. According to the Bahá'í faithful, Bahá'u'lláh appeared as a successor to God's previous great Earth-based Messengers, or Manifestations, among them Krishna, Buddha, Moses, Zoroaster, Jesus and Muhammad.

Central to the Bahá'í Faith are the ideas that each human has a non-material soul created by God and that the physical body of each human being offers a temporary home for that human's soul. According to the beliefs of the Bahá'í Faith, both the soul and that soul's human body are created by God at the same moment. During one's brief lifetime, one seeks to fulfill "the purpose which that Creator has fixed for His creatures,"[1] which is "to know his Creator and to attain His Presence."[2] After the physical death of the body, the soul endures in the spiritual world, which exists outside of time and space. The soul is not reincarnated into another body in any other location in our physical universe, nor is any place in the physical universe the location of human souls in the afterlife.

Several decades after founding the religion now known as the Bahá'í Faith, Bahá'u'lláh appointed his eldest son 'Abdu'l-Bahá (1844–1921 C. E.) as his spiritual successor and

---

[1] Retrieved from http://info.bahai.org/article-1-4-0-6.html

[2] Bahá'u'lláh. (1990). *Gleanings from the writings of Bahá'u'lláh* (pocket-size edition, p. 70). US Bahá'í Publishing Trust. Retrieved from http://reference.bahai.org/en/t/b/GWB/

D.A. Weintraub, *Religions and Extraterrestrial Life: How Will We Deal With It?*,
Springer Praxis Books, DOI 10.1007/978-3-319-05056-0_20,
© Springer International Publishing Switzerland 2014

leader of the faith. 'Abdu'l-Bahá was given the authority to be the sole interpreter of the writings of Bahá'u'lláh and the respect from the Bahá'í Faith community that his interpretations would be accepted as theological absolutes. "Bahá'u'lláh, in the Book of His Covenant, confirmed the appointment of His Son 'Abdu'l-Bahá as the interpreter of His Word and the Center of His Covenant. As the interpreter, 'Abdu'l-Bahá became the living mouth of the Book, the expounder of the Word; as the Center of the Covenant, He became the incorruptible medium for applying the Word to practical measures for the raising up of a new civilization."[3] Since some of the words and writings of Bahá'u'lláh and the interpretations thereof by 'Abdu'l-Bahá are about science and extraterrestrial life, the words and writings of 'Abdu'l-Bahá provide a vehicle for understanding how the Bahá'í would handle knowledge about extraterrestrials.

One of the fundamental tenets of the Bahá'í Faith is the unity of religion and science. For the Bahá'í faithful, religion and science are systems of knowledge that must be in harmony. Religion must withstand the analysis of reason. What is learned from one must reinforce what is learned from the other, and no conflict can exist between the two. As explained by 'Abdu'l-Bahá, "Bahá'u'lláh has declared that religion must be in accord with science and reason. If it does not correspond with scientific principles and the processes of reason, it is superstition. For God has endowed us with faculties by which we may comprehend the realities of things, contemplate reality itself. If religion is opposed to reason and science, faith is impossible; and when faith and confidence in the divine religion are not manifest in the heart, there can be no spiritual attainment."[4]

The Bahá'í also strongly espouse expanding our knowledge about the universe. Doing so, according to 'Abdu'l-Bahá, is a religious imperative: "There are certain pillars which have been established as the unshakeable supports of the Faith of God. The mightiest of these is learning and the use of the mind, the expansion of consciousness, and insight into the realities of the universe and the hidden mysteries of Almighty God. To promote knowledge is thus an inescapable duty imposed on every one of the friends of God."[5]

In his 1945 tract *Foundations of World Unity*, 'Abdu'l-Bahá writes that "as God is creator, eternal, and ancient, there were always creatures and subjects existing and provided for. There is no doubt that divine sovereignty is eternal. ... If we conceive of a time when there were no creatures ... we dethrone God and predicate a time when God was not."[6] The logic presented by 'Abdu'l-Bahá is straightforward: God has always existed—God "is eternal"—and living creatures created by God have always existed—"there were always creatures"—in God's universe. However, since Earth and humankind on Earth are

---

[3] "Message on the Centennial Day of the Covenant" (1992, November 26), in *Letters from the Universal House of Justice, 1986–2001*; Retrieved from http://en.bahaitext.org/MUHJ86-01/145/Message_on_the_Centennial_of_the_Day_of_the_Covenant

[4] 'Abdu'l-Bahá. (1982). *The promulgation of universal peace* (2nd ed., pp. 298–299). US Bahá'í Publishing Trust. Retrieved from http://reference.bahai.org/en/t/ab/PUP/

[5] 'Abdu'l-Bahá. (1982). In *selections from the writings of 'Abdu'l-Bahá* (Bahá'í World Centre: Haifa), section 97 [19]; Retrieved from http://reference.bahai.org/en/t/c/SCH/

[6] 'Abdu'l-Bahá. (1945). *Foundations of world unity* (p. 101). Wilmette, IL: US Bahá'í Publishing Trust. Retrieved from http://reference.bahai.org/en/t/c/FWU

recent creations of God, this statement of 'Abdu'l-Bahá very clearly establishes that the Bahá'í believe that life must exist beyond the Earth.

Furthermore, as is clear from additional texts, the Bahá'í believe that at least some of those extraterrestrial beings are also humans. Humans have always existed, even though humankind on Earth is a recent addition to the universe. The faith-based evidence for this claim is as follows. According to the Bahá'í Faith, the progressive revelations of God continue in a process involving a Manifestation, i.e., a prophet. The Bahá'í identify nine Earth-based Manifestations, including both the Báb (a Persian born in 1819 C. E. who on May 23, 1844, "announced the imminent appearance of the Messenger of God awaited by all the peoples of the world,"[7] that person being Bahá'u'lláh) and Bahá'u'lláh. The Bahá'í assert, furthermore, that "Thus there have been many holy Manifestations of God. One thousand years ago, two hundred thousand years ago, one million years ago, the bounty of God was flowing, the radiance of God was shining, the dominion of God was existing."[8] But even before the Earth existed, Perfect Manifestations did exist, as did humankind: "All that we can say is that this terrestrial globe at one time did not exist, and at its beginning man did not appear upon it. But from the beginning what has no beginning, to the end which has no end, a Perfect Manifestation always exists. … Therefore, it cannot be imagined that the worlds of existence, whether the stars or this earth, were once inhabited by the donkey, cow, mouse and cat, and that they were without man! This supposition is false and meaningless."[9]

Perhaps the most well-known statement made by 'Abdu'l-Bahá that affirms with absolute certitude the Bahá'í view that extraterrestrial life exists is the following:

"The earth has its inhabitants, the water and the air contain many living beings and all the elements have their nature spirits, then how is it possible to conceive that these stupendous stellar bodies are not inhabited? Verily, they are peopled, but let it be known that the dwellers accord with the elements of their respective spheres. These living beings do not have states of consciousness like unto those who live on the surface of this globe: the power of adaptation and environment moulds their bodies and states of consciousness, just as our bodies and minds are suited to our planet."[10] This assertion by 'Abdu'l-Bahá is consistent with statements by many others in the late nineteenth and early twentieth centuries, when belief in extraterrestrial life was common and the principle of plenitude remained widely accepted.

Bahá'u'lláh himself, sounding very much like a nineteenth-century version of Giordano Bruno, wrote on page 163 in his *Gleanings from the Writings of Bahá'u'lláh* that the universe is full of an uncountable number of stars orbited by inhabited planets: "Thou hast, moreover, asked Me concerning the nature of the celestial spheres. To comprehend

---

[7] From info.bahai.org/the-bab.html

[8] 'Abdu'l-Bahá. (1982). *The promulgation of universal peace*, "4 December 1912 Talk to the Theosophical Society," *Peace* (2nd ed., p. 463). Wilmette, IL: US Bahá'í Publishing Trust. Retrieved from http://reference. bahai.org/en/t/ab/PUP

[9] 'Abdu'l-Bahá. (1981). *Some answered questions* (5th ed., pp. 196–197). Wilmette, IL: US Bahá'í Publishing Trust. Retrieved from http://reference.habai.org/en/t/ab/SAQ/

[10] 'Abdu'l-Bahá. (1918). *Abdul Baha on Divine Philosophy*, authors 'Abdu'l-Bahá and Isabel Fraser Chamberlain (Bahai Publishing Committee, pp. 114–115).

their nature, it would be necessary to inquire into the meaning of the allusions that have been made in the Books of old to the celestial spheres and the heavens, and to discover the character of their relationship to this physical world, and the influence which they exert upon it. Every heart is filled with wonder at so bewildering a theme, and every mind is perplexed by its mystery. God, alone, can fathom its import. The learned men, that have fixed at several thousand years the life of this earth, have failed, throughout the long period of their observation, to consider either the number or the age of the other planets. Consider, moreover, the manifold divergencies that have resulted from the theories propounded by these men. Know thou that every fixed star hath its own planets, and every planet its own creatures, whose number no man can compute."[11]

'Abdu'l-Bahá, in continuing the tradition and will of Bahá'u'lláh in appointing the first-born of his lineal male descendents as his successor, appointed his eldest grandson, Shoghi Effendi (1897–1957 C. E.), to succeed him as the Guardian of the Bahá'í Faith, to be "the Interpreter of the Word of God."[12] Effendi, in a letter written in 1937 while acting in his role as Guardian of the Faith, provided a Bahá'í answer to the question "What will extraterrestrials be like?" when he interpreted these words of Bahá'u'lláh in this way: "Regarding the passage on p.163 of the 'Gleanings'; the creatures which Bahá'u'lláh states to be found in every planet cannot be considered to be necessarily similar or different from human beings on this earth. Bahá'u'lláh does not specifically state whether such creatures are like or unlike us. He simply refers to the fact that there are creatures in every planet. It remains for science to discover one day the exact nature of these creatures."[13] Shoghi Effendi also wrote that though we know little about the nature of these creatures, we can be certain that they are intelligent and can know God: "'Abdu'l-Bahá stated there are other worlds than ours which are inhabited by beings capable of knowing God."[14]

Many of the writings of Bahá'u'lláh were collected and published in the *Tablets of Bahá'u'lláh Revealed After the Kitáb-i'Aqdas*. In the tablet known as Súriy-i-vafá, he made very clear that God had made unnumerable worlds, many or most with living beings: "As to thy question concerning the worlds of God, Know thou of a truth that the worlds of God are countless in number, and infinite in their range. None can reckon them except God … Verily I say, the creation of God embraceth worlds besides this world, and creatures apart from these creatures."[15]

Bahá'u'lláh had written earlier about whether extraterrestrials would follow his prophetic teachings or discover their own path to knowing God. He concluded that each

---

[11] Bahá'u'lláh. (1990). *Gleanings from the writings of Bahá'u'lláh* (pocket-size edition, p.162–163). US Bahá'í Publishing Trust. Retrieved from http://reference.bahai.org/en/t/b/GWB/

[12] *The Bahá'í World, Volume 7, Part 2;* Retrieved from http://en.bahaitext.org/The_Bahá'í_World/Volume_VII/Part_II

[13] Effendi, S. (1988). Letter of February 9, 1937, quoted in *Lights of Guidance*, Helen Hornby, #1581 (2nd ed.). New Delhi: Bahá'í Publishing Trust. Retrieved from http://en.bahaitext.org/Lights_of_Guidance/Explanation_of_Some_Bahá'í_Teachings

[14] Effendi, S. *The light of divine guidance* (Vol. II, p. 95). Retrieved from http://reference.bahai.org/en/t/se/LDG2/ldg2-95.html

[15] Bahá'u'lláh. *Tablets of Bahá'u'lláh Revealed After the Kitáb-i'Aqdas: Súriy-i-vafá* (p. 188). Retrieved from http://reference.bahai.org/en/t/b/TB-tb-13

inhabited world will receive its own prophet: "O people! I swear by the one true God! This is the Ocean out of which all seas have proceeded, and with which every one of them will ultimately be united. From Him all the Suns have been generated, and unto Him they will all return. Through His potency the Trees of Divine Revelation have yielded their fruits, every one of which hath been sent down in the form of a Prophet, bearing a Message to God's creatures in each of the worlds whose number God, alone, in His all-encompassing Knowledge, can reckon. This He hath accomplished through the agency of but one Letter of His Word, revealed by His Pen—a Pen moved by His directing Finger—His Finger itself sustained by the power of God's Truth."[16]

If the inhabitants of each world worship God through their world's own prophetically-revealed religion, then we can understand the God worshiped by the Bahá'í faithful as a universal God but the Bahá'í Faith itself as a religion only for humankind. Shoghi Effendi explained in 1938 in a letter to a member of the faithful, however, that the spirit and scope of the Revelation of Bahá'u'lláh could extend beyond the Earth: "As to your question whether the power of Bahá'u'lláh extends over our solar system and to higher worlds; while the Revelation of Bahá'u'lláh, it should be noted, is primarily for this planet, yet the spirit animating it is all-embracing, and the scope therefore cannot be restricted or defined."[17]

We can take one further step in understanding whether the Bahá'í Faith might truly be limited to humans on Earth when we take note of 'Abdu'l-Bahá's remarks to the National Association for the Advancement of Colored People in 1912. In these remarks, he made clear that while all intelligent creatures are made "in the image and likeness of God," the Bahá'í interpret "image and likeness of God" in terms of "intelligence and spirit," not color and not, we can presume, shape, form, texture, size or any other physical attributes: "Let us now discover more specifically how he is the image and likeness of God and what is the standard or criterion by which he can be measured and estimated. This standard can be no other than the divine virtues, which are revealed in him. Therefore, every man imbued with divine qualities, who reflects heavenly moralities and perfections, who is the expression of ideal and praiseworthy attributes, is, verily, in the image and likeness of God. If a man possesses wealth, can we call him an image and likeness of God? Or is human honor and notoriety the criterion of divine nearness? Can we apply the test of racial color and say that man of a certain hue—white, black, brown, yellow, red—is the true image of his Creator? We must conclude that color is not the standard and estimate of judgment and that it is of no importance, for color is accidental in nature. The spirit and intelligence of man is essential, and that is the manifestation of divine virtues, the merciful bestowals of God, the eternal life and baptism through the Holy Spirit. Therefore, be it known that color or race is of no importance. He who is the image and likeness of God, who is the manifestation of the bestowals of God, is acceptable at the threshold of

---

[16] Bahá'u'lláh. (1990). *Gleanings from the writings of Bahá'u'lláh* (pocket-size edition, p. 104). US Bahá'í Publishing Trust. Retrieved from http://reference.bahai.org/en/t/b/GWB/

[17] Effendi, S. (1988). Letter of July 14, 1938, quoted in *Lights of Guidance*, Helen Hornby, #1594 (2nd ed.). New Delhi: Bahá'í Publishing Trust. Retrieved from http://en.bahaitext.org/Lights_of_Guidance/Explanation_of_Some_Bahá'í_Teachings

God—whether his color be white, black or brown; it matters not. Man is not man simply because of bodily attributes. The standard of divine measure and judgment is his intelligence and spirit."[18]

If, in fact, the only common denominator across the universe among intelligent beings is intelligence and spirit, it is unlikely that the prophets to those other worlds would be worried about many of the basic principles of the Bahá'í Faith, which include the important twentieth and twenty-first century issues of unifying our global society, the abandonment of prejudice, the equality of opportunities for men and women, the elimination of poverty and wealth, and universal education, all of which appear to be very specific to the lives of humans living on Earth. Almost certainly, then, the Bahá'í Faith is a religion only for human beings located on planet Earth. Extraterrestrials, the Bahá'í faithful would expect, will worship the same God in their own world-specific way, because, as Bahá'u'lláh writes in *Gleanings*, "All glory be to God, the Lord of all worlds."[19]

[18] 'Abdu'l-Bahá. (1982). *The promulgation of universal peace* (2nd ed., pp. 69–70). US Bahá'í Publishing Trust. Retrieved from http://reference.bahai.org/en/t/ab/PUP/

[19] Bahá'u'lláh. (1990). *Gleanings from the writings of Bahá'u'lláh* (pocket-size edition, p.172). US Bahá'í Publishing Trust. Retrieved from http://reference.bahai.org/en/t/b/GWB/

21

Are We Ready?

> *Kaffee: I want the truth!*
> *Jessup: You can't handle the truth!*
>
> An exchange between Lieutenant Kaffee (played by Tom
> Cruise) and Colonel Jessup (played by Jack Nicholson) in
> the movie <u>A Few Good Men</u> (1992)

THE RELATIONSHIP BETWEEN ASTRONOMY AND RELIGION

Astronomy and religion have been intimate partners for far longer than humans have been recording their histories. Each and every day, ancient peoples needed to find enough food to nourish their bodies. Each and every evening by sundown, they needed to shelter themselves in secure places where they could hope to sleep in safety through the perilous hours of darkness. At sunrise, they would arise and work again to obtain sustenance for themselves. Another sunset followed by another sunrise meant another day still alive, another day during which mothers could give birth to babies, children could grow bigger and stronger, and young adults could be taught the ways of the tribe so that they might lead their communities through future days and nights.

One of the first major discoveries about the natural world made by our ancient forebears was an understanding of the cycle of the seasons. They quickly recognized how to associate the numbers of hours of sunshine per day and the intensity of sunlight each day with the waxing and waning of the seasons, the ripening of fruits and nuts and the birth and growth of game animals. From such associations came recognition of causation: the Sun controlled their health and survival. With help from the Sun, food was plentiful and people were warm. Without enough sunlight, people starved or froze. People living along coastlines learned that the Moon controlled the ebb and flow of the tides. Understanding tidal patterns allowed them to hunt and collect the bounty of the sea successfully at the best

D.A. Weintraub, *Religions and Extraterrestrial Life: How Will We Deal With It?*,
Springer Praxis Books, DOI 10.1007/978-3-319-05056-0_21,
© Springer International Publishing Switzerland 2014

times of each month and during the best seasons of each year. They learned how to predict the onset of coming seasons and the cycle of the tides.

By observing and recognizing the annual pattern of changes in the location of the rising and setting Sun on the horizon (further to the north in summer; further to the south in winter) or the height of the midday Sun above the southern horizon (higher in summer; lower in winter) or the lengths of shadows (longer in winter than in summer; lengthening until mid-winter and shrinking until mid-summer) they could measure and predict the arrival of the seasons and plan for upcoming changes in weather patterns and in the availability of food. More advanced cultures would have discovered they could make the same predictions for the warming and cooling of the air and the waxing and waning of the seasons by observing stars in the nighttime sky and noting which stars appear in the sky each night or rise and set at a particular time of night at different times of year. Given that their very survival was at stake, our ancestors of long ago must have quickly learned to associate a sense of the divine with the bright, objects in the sky that appeared to have significant control over their health and welfare, over whether they lived or died. The Sun, the Moon, and the stars all became gods whom ancient peoples believed controlled the destinies of individuals and of entire human societies. At this point in the cultural development of humankind, many tens of thousands of years ago, humanity established a clear and overwhelmingly important link between astronomy and religion.

For thousands of years, astronomers have played an extremely significant, even if often hidden, role in religious practice, primarily by tracking time and establishing and maintaining calendars. The Sun, after all, gives us our day; the Sun and stars give us our year; the Moon gives us our weeks and months. As Plato wrote in his *Timaeus*, the creator of the universe sought to make the universe eternal. In order to do so, he set in the heavens "a moving image of eternity. ... The sun and moon and five other stars, which are called the planets, were created by him in order to distinguish and preserve the numbers of time."[1]

A similar tale is told in the familiar words of Genesis: "And God said, Let there be lights in the firmament of the heaven to divide the day from the night; and let them be for signs, and for seasons, and for days, and years: And let them be for lights in the firmament of the heaven to give light upon the earth: and it was so. And God made two great lights; the greater light to rule the day, and the lesser light to rule the night: he made the stars also. And God set them in the firmament of the heaven to give light upon the earth, And to rule over the day and over the night, and to divide the light from the darkness: and God saw that it was good. And the evening and the morning were the fourth day."[2]

Since the birth of Islam, astronomers have played a vital role in the practice of Islam by identifying the direction of Mecca from any location on the surface of the Earth (not an easy task on the surface of a spherical planet, especially if done from a moving ship in the middle of an ocean) and establishing the correct times of day for prayer. In addition, each month the Islamic calendar begins with the first sighting of the newborn, crescent moon, and trained observers, if not astronomers, are of vital importance for watching for this event and reporting as to the moment when it occurs.

---

[1] Plato. *Timaeus*. (B. Jowett, trans.). Retrieved from http://classics.mit.edu/Plato/timaeus.html)

[2] Genesis 1, 14:19, King James Version.

In Christianity, the date of Easter is set as the first Sunday after the first full moon after the vernal equinox in the northern hemisphere. The vernal equinox is commonly known as the first day of northern Spring, when the lengths of the day and night are nearly the same (about 12 h) at all latitudes on the Earth. More correctly, in the northern hemisphere the vernal equinox is the exact moment when the center of the Sun crosses the celestial equator, moving from south to north.

Twice, astronomers have played an all-important role in correcting the western calendar. First, in 46 B. C. E. Julius Caeser, on advice from astronomer Sosigenes, added the leap year to the Roman calendar, thereby putting in place the so-called Julian Calendar. Then, just over 1,600 years later, Pope Gregory XIII, who relied on advice from astronomer Christoph Clavius and the astronomers in the Collegio Romano in Rome, fixed the western calendar so that Easter and Christmas would continue to be celebrated in the spring and the winter, respectively.

With the geography seemingly well understood and the calendar well managed and now under the control of secular authorities (the International Earth Rotation and Reference Systems Service determines when leap seconds should be added to atomic clocks around the world), the obvious links between astronomy and religion have weakened in recent centuries. Nevertheless, the questions astronomers continue to ask about the universe are similar to questions theologians still ask: Did the universe have a beginning? How does the Earth fit into the universe? Are humans alone in the universe? Now, astronomers are poised to play perhaps their most significant role in human history, not just in human religious history, as the discoveries of and about exoplanets they may make in this century could shatter the worldviews of much of humanity.

ARE WE READY?

If, tomorrow, humanity learned that astronomers had unequivocal evidence that life exists on another planet or that the Secretary General of the United Nations had been conversing with little green beings who had landed their flying saucer on the Champs-Elysées in Paris, what would we do? What would you do?

Could we handle the truth? Are we ready to embrace the idea that we are not the only intelligent beings and perhaps not even the most intelligent or technologically advanced creatures in the universe? Are we prepared for the knowledge that the Earth is not the sole host for living organisms in the universe? Would that knowledge have any impact on our relationships with each other?

While some of us claim to be ready, a great many of us probably are not. If opinion surveys accurately probe and portray our collective views on the matter, most of us are sure extraterrestrials exist while only a small minority among us are equally certain they do not. From a scientific perspective, however, the only thing we know about extraterrestrials today with one-hundred percent certainty is that none of us know whether humanity is alone or has company in the universe. One other thing is assuredly true: very few among us have spent much time thinking hard about what actual knowledge about extraterrestrial life, whether viruses or single-celled creatures or bipeds piloting intergalactic spaceships, might mean for our personal beliefs, our relationships with the divine, and our collective place, role and responsibilities to and for life here on Earth as well as elsewhere in the universe.

While actual contact with intelligent Others likely remains an event that lies far in the future, it might not. Knowledge of the existence of extraterrestrial life could easily arrive via the light astronomers collect with their telescopes as soon as tomorrow or next year. Given how profound that knowledge would be, should we begin talking amongst ourselves about what knowing that we are not alone in the universe would mean for us?

In the religious realm, most religions appear theologically open to the possibility that intelligent extraterrestrial beings exist. Most religious belief systems are robust enough to accommodate the paradigm-busting news that the discovery of extraterrestrial life would represent. For those who identify with one of these religious traditions, the discovery of extraterrestrial life of any kind would likely confirm for them the awesome power of the God in whom they believe and provide affirmation to them of the truths offered through the teachings of their respective religions about our place in the universe. As for those who hold religious beliefs that deny the possibility that sentient beings beyond the Earth exist, those beliefs could soon be challenged. Historically, some religions have shown that they have enough theological dexterity to survive the regular challenges to doctrine and belief that emerge from humankind's increasingly sophisticated knowledge about the natural world (e.g., despite its very public spat with Galileo, the Roman Catholic Church has adapted to the scientific knowledge that the Earth orbits the Sun). Other religions (e.g, the Fates and the Greek pantheon) lacked the necessary doctrinal flexibility to survive.

Even if, in the next few centuries, humanity does not uncover any evidence of extra-terrestrial life, we should remember the maxim that the absence of evidence is not the evidence of absence. With this idea in mind, contemplating the possible reality of extrater-restrial Others, even without any evidence (yet) that they exist, might nevertheless be an activity worthy of our time and energy.

What might change if we woke up to news that such Others had been identified? The belief structures for Jews, Roman Catholics, Quakers, Seventh-day Adventists and mem-bers of the Bahá'í Faith should allow them to accept the reality of extraterrestrials without having to deny their own religious beliefs; the faith that Buddhists, Hindis, Sikhs and Jains have that the God in whom they have faith is the God of the entire universe should allow them to accept that sentient beings from other worlds can share in the grace of this univer-sal God without the Others having to convert to Buddhism, Hinduism, Sikhism or Jainism; and the logic built within their faiths for Muslims, Methodists and Mormons that their religions are meant only for sentient beings who are residents of Earth and that their prophets spoke words meant only for Earthlings should permit most of them to accept that extraterrestrials might have their own religions and prophets. In contrast, many Orthodox Christians and fundamentalist and evangelical Christians may struggle to reconcile their beliefs with news that extraterrestrial life exists, especially if the extraterrestrials are intel-ligent and more technologically advanced than humanity.

If most of us can figure out how to reconcile our religious beliefs with the existence of extraterrestrial intelligences who do not share our religious beliefs and practices, and if most of us can recognize that our religions only make sense on Earth, that our religious beliefs and practices are terrestrial and not universal, even if the God in whom one believes is worshiped as the God of the entire universe, perhaps we can learn a lesson from thinking about how we would interact with and understand extraterrestrial Others. If our understanding of our own religions allow us to live peacefully with those from other worlds whose religions are likely different from ours, then perhaps we can learn to live more peaceably with 'terrestrial intelligences' with whom we are already well acquainted and whose religious beliefs and practices differ from our own.

REVISING THE COSMOLOGICAL PRINCIPLE

Astronomers often refer to the cosmological principle in this form: *We do not live in a special place in the universe.* The idea that we do not occupy a special place in the universe is a relatively new idea for most of humanity. Once upon a time we thought we did live in a unique place in the universe. Most of humanity thought the Earth was the center of the knowable universe, and the center of the universe must be a very special place. Many believers of many faiths found comfort and reassurance knowing that humans lived in the singularly most special location in the universe. Clearly, if we live at the center of the universe, the universe is necessarily centered on us both physically and spiritually. But in 1543 C. E., when Copernicus set the Earth in motion around the Sun, he simultaneously moved the center of the universe 150 million kilometers from the Earth to the Sun and gave birth to the cosmological principle. Thanks to Copernicus, the Earth became just one of the then-known six planets orbiting the Sun, no longer quite as special.

More recent discoveries in astronomy have further lessened the apparent specialness of the Earth and Sun, at least in terms of the physical location of the solar system in the universe. In the second decade of the twentieth century, astronomers proved that the Sun was a few tens of thousands of light years from the center of the Milky Way, and the Sun was just one of hundreds of billions of stars in the Milky Way. For a period of just a few years, most astronomers were convinced that the Milky Way was the entire universe. In the late 1920s, however, Edwin Hubble discovered that the mysterious celestial objects known as spiral nebulae were other Milky Ways, other galaxies, each containing tens or hundreds of billions of stars. In the 1930s, astronomers learned that the Milky Way was but one among billions of galaxies in an expanding universe that has no edge and no center.

What's next? Astronomers in the twenty-first century may soon prove that the universe is infinite in size. All the knowledge about the universe obtained by astronomers in recent decades has added to our certainty that we do not live in a special place in the universe. The Earth is a typical planet orbiting a typical star in a typical galaxy in a typical part of the universe. The laws of physics, according to the cosmological principle, must be the same throughout the universe; the materials that are the fundamental building blocks of all of the galaxies and stars and planets are the same materials—hydrogen, helium, carbon, oxygen and the other elements composed of protons, neutrons and electrons— found everywhere in the universe. If the contents and the laws of physics are the same everywhere in the universe, then galaxies and stars and planets should form in all parts of the universe, and the chances that life might arise—whether by chance or by the creative powers of a deity—are the same throughout the universe.

While we may not live in a special place in the universe, we may, however, live during a *very special time*: *some of us may be alive when astronomers discover life beyond the Earth.* Given the high likelihood that this discovery could occur soon, perhaps we should seriously consider beginning to prepare ourselves philosophically and theologically for this coming change in how we will understand ourselves and our place in the universe.

Appendix: The Exoplanets Revealed by the Kepler Mission Through 2013

> *In the beginning of the mission, no one knew whether Earth-size planets were abundant or rare in our galaxy. Now, with the completion of the Kepler observations, we know our galaxy is filled to the brim with planets.*
>
> William Borucki (August 15, 2013) (Harwood, W. CBS News Report. NASA's Kepler will no longer hunt for Alien planets. http://www.cbsnews.com/news/ nasas-kepler-will-no-longer-hunt-for-alien-planets/)

When on March 6, 2009 NASA launched the Kepler mission from Cape Canaveral into orbit around the Earth, the Kepler team, led by Principal Investigator William Borucki of NASA's Ames Research Center, catapulted the field of exoplanet science into the forefront of modern astronomical discovery work. By the end of 2013, Kepler had identified 167 exoplanets whose existence had been confirmed by careful follow-up studies using other telescopes. These planets include hot Jupiters, hot super-Earths, hot Earth-size planets, planets in habitable zones, multi-planet systems and planets orbiting double stars. Another 3,568 targets remain on Kepler's list of candidate planets and most of them likely will be revealed as bona fide planets within the next few years. These descriptions of the Kepler planets that have been discovered through the end of 2013 provide a robust introduction to astronomers' rapidly-growing library of knowledge about exoplanets.

The Kepler mission telescope simultaneously monitored the brightnesses of about 150,000 stars in the constellations of Cygnus and Lyra, most of which are located within the spiral arm of our Milky Way galaxy known as the Orion arm. Most of the Kepler target stars lie at distances of at least a few hundred light years from Earth. Designed as a transit detection telescope, Kepler searched for planets that passed in front of and dimmed the light of their host stars. Kepler recorded a measurement of the brightness of every star in its study sample every 30 min, and repeated these measurements continuously for about 4 years. Mechanical problems ended Kepler's primary mission in early 2013, though the Kepler data will continue to be analyzed by astronomers for years to come.

D.A. Weintraub, *Religions and Extraterrestrial Life: How Will We Deal With It?*, Springer Praxis Books, DOI 10.1007/978-3-319-05056-0,
© Springer International Publishing Switzerland 2014

For Kepler, as for any transit detection telescope, big planets close to their host stars are the easiest to detect because they orbit quickly—thereby producing more transit events during the course of Kepler's 4-year observing program—and block out a larger fraction of the light of the host star than would a smaller planet; however, Kepler is capable of detecting planets as small as the Earth in Earth-like (several hundred day) orbits. Such an Earth-size planet in an Earth-like orbit would dim the light from the central star by about 1 part out of 10,000 (that is the total light from the star would drop from 100 to 99.99 %) for about 12 h every 400 days.

THE FIRST-DISCOVERED KEPLER PLANETS

With only Kepler data, astronomers can determine the orbital period and the physical size (the radius or diameter) of a planet. To know what a particular planet is made of (the composition), they also need to measure the mass of the planet, but that measurement must be made with a different telescope. In combination, the mass and the diameter yield the average density of the planet, and the density of a planet allows astronomers to compare its composition to those of the Sun, the Earth and the other planets in the solar system. Astronomers will then compare the measured density of the exoplanet to the known densities of the Sun and the planets in our solar system. Planets like Jupiter (average density 1.3 g/cm^3) and Saturn (0.69) have very low densities, similar to that of the Sun (1.41). This information about their densities tell us that they, like stars, are made predominantly of the lightest elements, the gases hydrogen and helium. If we could place Saturn in Earth's oceans, it would float like an ice cube because its density is less than the density of water (1.00 g/cm^3 at sea level on the Earth). In contrast, Mercury (5.42), Venus (5.42), Earth (5.52), Mars (3.94) and the Moon (3.34) have much higher densities, because they are made predominantly of silicate rocks (composed of silicon and oxygen, along with assorted other elements like calcium, magnesium and aluminum) and metals (mostly iron and nickel).

For the sake of comparisons, let's define several size categories for exoplanets:

- The Jupiters and super-Jupiters have radii comparable to or larger than the radius of Jupiter (71,400 km) and likely have masses of about half to about 12 times the mass of Jupiter. For comparison, the radius of Jupiter is 11 times greater than the radius of the Earth, and the mass of Jupiter is equivalent to 318 Earth masses;
- The Saturns and super-Saturns have radii comparable to or larger than the radius of Saturn (60,330 km) and have masses ranging from about 50 to about 150 Earth masses (the mass of Saturn is 95 Earth masses);
- The Uranus-like and Neptune-like exoplanets and super-Neptunes are similar to or bigger than Uranus (radius 25,579 km, or 4.0 Earth radii; 15 Earth masses) and Neptune (radius 24,764 km, or 3.9 Earth radii; 18 Earth masses), but smaller than Saturn-like exoplanets; all exoplanets with radii greater than twice the radius of the Earth typically belong in the Uranus-like and Neptune-like categories; planets like Uranus and Neptune are made mostly of materials astronomers call ices—these being frozen water, methane and ammonia;

- Super-Earths have radii between 1.25 and 2.0 Earth radii; while their masses are usually unknown, their masses are likely in the range of two to eight Earth masses; planets at the smaller end of this mass range might be like the Earth, mostly made of rock and with solid surfaces; planets at the high end of this mass range are more likely ice-giants like Uranus and Neptune or gas-giants like Saturn and Jupiter;
- Earth-like planets have radii smaller than 1.25 Earth radii; planets of this size are almost certainly rocky, iron-rich planets with solid surfaces; if the size of their orbits are larger than a few tenths of an astronomical unit, these planets likely have thin atmospheres above their solid surfaces.

Kepler scientists launched their era of big discoveries when in late 2009, with only a few months of data in hand, they announced that they had discovered five giant planets, all with orbits of less than 5 days. None of these broke the mold of the hot Jupiters discovered with ground-based telescopes, but they quickly demonstrated that detecting exoplanets with Kepler would be, relatively speaking, easy pickings. Borucki, Deputy Principal Investigator David Koch, also at NASA's Ames Research Center, and team Co-Investigators Ted Dunham, of Lowell Observatory, Jon Jenkins, of the SETI Institute in Mountain View, CA, and David Latham, of the Harvard-Smithsonian Astrophysical Observatory, shared credit with dozens of other Kepler-team scientists in announcing these first five discoveries.

- **Kepler-4b** is the smallest of these first five, being Neptune-like in mass (24 Earth masses), radius (4 Earth radii), and density (1.9 g/cm^3; Neptune's density is 1.64 g/cm^3). At a distance of only 0.0456 astronomical units from the central star known as Kepler-4, Kepler-4b is 21 times closer to its star than is the Earth to the Sun, orbits Kepler-4 in only 3.2 days, and is extremely hot, with a temperature of about 1,700 K.[1]
- **Kepler-5b** is quite a bit bigger than Kepler-4b, being 2.1 times more massive than Jupiter (or 670 times more massive than the Earth) and 1.4 times bigger in radius than Jupiter (or almost 16 times bigger in radius than the Earth). Despite its large mass, because of its extreme girth Kepler-5b has a very low density and, like Saturn, would float in water. Like Kepler-4b, Kepler-5b is very close to its host star (0.051 astronomical units) and as a result is also extremely hot (1,800 K).[2]
- **Kepler-6b** is a mini-Jupiter in mass (0.7 Jupiter masses) but a maxi-Jupiter in size (1.3 Jupiter radii). Consequently it is even less dense than Kepler-5b. With an orbit of 3.2 days at a distance of 0.046 astronomical units from the central star, Kepler-6b is also extremely hot, with a temperature of 1,500 K.[3]
- **Kepler-7b** is a smaller version of Kepler-6b, making it more like Saturn than like Jupiter. With a mass of 0.4 Jupiters (or 1.3 Saturns) and a radius of 1.6 Jupiters (or 1.9 Saturns), Kepler-7b has a density less than one-fifth that of water. Like these others of

[1] Borucki, J. W., et al. (2010). Kepler-4b: A hot Neptune-like planet of a G0 star near main-sequence turn-off. *Astrophysical Journal, 713*, L126.

[2] Koch, D. G., et al. (2010). Discovery of the transiting planet Kepler-5b. *Astrophysical Journal, 713*, L131.

[3] Dunham, E. W., et al. (2010). Kepler-6b: A transiting hot Jupiter orbiting a metal-rich star. *Astrophysical Journal, 713*, L136.

Kepler's first discoveries, Kepler-7b is in a very small (0.06 astronomical units) and quick (4.9 days) orbit and consequently is also very hot (1,500 K).[4]

- The last of the first five discoveries is **Kepler-8b,** which is most similar to Kepler-6b and Kepler-7b (0.6 Jupiter masses, 1.5 Jupiter radii, a density intermediate between those of Kepler-6b and Kepler-7b), and like the others, it is in the class of hot Jupiters (orbital size 0.048 astronomical units; orbital period 3.5 days; temperature 1,800 K).[5]

Data from Kepler has revealed so many hot Jupiters and super-Jupiters and hot Neptunes and super-Neptunes that the discovery of objects like these has become almost commonplace. Astronomers normally publish their results in refereed professional journals and then hope that the popular press will pick up the story. The *New York Times*, which has published the news of many planet discoveries, now routinely ignores the discovery-related press releases for most hot Jupiters, as these discoveries apparently are no longer sufficiently newsworthy to be included in "All the News That's Fit to Print."

The hot Jupiters recently discovered by Kepler include:

- **Kepler-12b**, a 0.43 Jupiter-mass planet in a 4.438-day orbit, orbiting at a distance of only 0.0556 astronomical units, with a 'super-inflated radius' of 1.7 Jovian radii[6];
- **Kepler-13b**, a planet in a 1.76-day orbit in a triple-star system, that is, a star system in which a single star and a binary star system orbit each other; in this system, Kepler-13b orbits the single star, which also happens to be the brightest and most massive star in this system; the mass of the planet has been determined to be 8.3 Jupiter masses by one observing team and no greater than 14.8 Jupiter masses by another[7];
- **Kepler-14b**, a 8.4 Jupiter-mass planet in a 6.79-day orbit, orbiting at a distance on 0.079 astronomical units[8];
- **Kepler-15b**, a 0.66 Jupiter-mass planet in a 4.94-day orbit, orbiting at a distance on 0.0571 astronomical units, with as much as 20 % of the mass composed of rocky and iron materials (compared to only about 3 % in Jupiter)[9];
- **Kepler-17b**, a 2.45 Jupiter-mass planet in a 1.49-day orbit, orbiting at a distance on 0.0259 astronomical units[10];

[4]Latham, D. W., et al. (2010). Kepler-7b: A transiting planet with unusually low density. *Astrophysical Journal, 713*, L140.

[5]Jenkins J. M., et al. (2010). Discovery and Rossiter-McLaughlin effect of exoplanet Kepler-8b. *Astrophysical Journal, 724*, 1198.

[6]Fortney, J. J., et al. (2011). Discovery and atmospheric characterization of giant planet Kepler-12b: An inflated radius outlier. *Astrophysical Journal Supplement Series, 197*, 9.

[7]Mislis, D., & Hodgkin, S. (2012). A massive exoplanet around KOI-13: Independent confirmation by ellipsoidal variations. *Monthly Notices of the Royal Astronomical Society, 422*, 1512; Santerne, A., et al. (2012). SOPHIE velocimetry of Kepler transit candidates VI. An additional companion in the KOI-13 system. *Astronomy & Astrophysics, 544*, L12.

[8]Buchhave L. A., et al. (2011). Kepler-14b: A massive hot Jupiter transiting an F Star in a close visual binary. *Astrophysical Journal Supplement Series, 197*, 3.

[9]Endl, M., et al. (2011). Kepler-15b: A hot Jupiter enriched in heavy elements and the first Kepler mission planet confirmed with the Hobby-Eberly telescope. *Astrophysical Journal Supplement Series, 197*, 13.

[10]Désert, J.-M., et al. (2011). The hot-Jupiter Kepler-17b: Discovery, obliquity from stroboscopic starspots, and atmospheric characterization. *Astrophysical Journal Supplement Series, 197*, 14.

- **Kepler-39b,** a 18 Jupiter-mass object (perhaps a planet, but perhaps sufficiently massive to be a small brown dwarf rather than a super-Jupiter) in a 21.1-day orbit[11];
- **Kepler-40b,** a 2.2 Jupiter-mass planet in a 6.87-day orbit around a very large (more than twice the radius of the Sun) host star[12];
- **Kepler-41b,** a 0.49 Jupiter-mass planet in a 1.86-day orbit, and with a day-side temperature of 1,900 K, circling a star almost identical to the Sun[13];
- **Kepler-43b,** a 3.23 Jupiter-mass planet in a 3.02-day orbit[14];
- **Kepler-44b,** a 1.02 Jupiter-mass planet in a 3.25-day orbit[15];
- **Kepler-45b,** a Jupiter-sized but only 0.50 Jupiter-mass planet in a 2.46-day orbit[16];
- **Kepler-76b,** a 2.0 Jupiter-mass planet in a 1.54-day orbit.[17]

The hot Neptunes recently discovered by Kepler include:

- **Kepler-19b,** a 2.2 Earth-radius planet of unknown mass, but one that is less massive than 14 Earths, orbiting at a distance of 0.1 astronomical units in only 9.287 days; however, the transits sometimes occur as much as 5 min early and other times as much as 5 min late, because a second planet is tugging on it[18];
- **Kepler-19c,** the second planet in the Kepler-19 system, known because it is gravitationally perturbing Kepler-19b, has a mass less than 31.6 Earths;
- **Kepler-21b,** a planet whose mass is less than 10 Earths and whose radius is 1.64 Earth radii; it orbits Kepler-21 in 2.79 days at a distance of 0.042 astronomical units and has a temperature of 1,960 K.[19]

---

[11] Bouchy, F., et al. (2011). SOPHIE velocimetry of Kepler transit candidates III. KOI-423b: An 18 M_{Jup} transiting companion around an F7IV star. *Astronomy & Astrophysics, 533,* A83.

[12] Santerne, A., et al. (2011). SOPHIE velocimetry of Kepler transit candidates. II. KOI-428b: A hot Jupiter transiting a subgiant F-star. *Astronomy & Astrophysics, 528,* A63.

[13] Santerne, A., et al. (2011). SOPHIE velocimetry of Kepler transit candidates IV. KOI-196b: A non-inflated hot-Jupiter with a high albedo. *Astronomy & Astrophysics, 536,* A70.

[14] Bonomo, A. S. et al. (2012). SOPHIE velocimetry of Kepler transit candidates V. The three hot Jupiters KOI-135b, KOI-204b, and KOI-203b (alias Kepler-17b). *Astronomy & Astrophysics, 538,* A96.

[15] Ibid.

[16] Johnson, J. A., et al. (2012). Characterizing the cool KOIs. II. The M Dwarf KOI0254 and its hot Jupiter. *Astronomical Journal, 143,* 111.

[17] Faigler, S., et al. (2013). BEER analysis of Kepler and CoroT light curves. I. Discovery of Kepler-76b: A hot Jupiter with evidence of superrotation. *Astrophysical Journal, 771,* 26.

[18] Ballard, S., et al. (2011). The Kepler-19 system: A transiting 2.2 R_{Earth} planet and a second planet detected via transit timing variations. *Astrophysical Journal, 743,* 200.

[19] Howell, S. B., et al. (2012). Kepler-21b: A 1.6 R_{Earth} planet transiting the bright oscillating subgiant star HD 179070. *Astrophysical Journal, 746,* 123.

MULTI-PLANET SYSTEMS

The first multi-planet system, identified by Kepler scientists in October 2010, is the system Kepler-9, now known to host at least three planets.[20]

- Matthew Holman, of the Harvard-Smithsonian Center for Astrophysics, led a team that determined that **Kepler-9b** is another super-Saturn, with a mass that is one-quarter the mass of Jupiter (84 % of the mass of Saturn), a radius that is 0.84 Jupiter radii (99 % the radius of Saturn), and an orbit just about one-third the size of the orbit of Mercury (0.14 astronomical units).
- **Kepler-9c** is less massive (57 % the mass of Saturn) but almost the same size (97 % the radius of Saturn) and in a slightly bigger orbit (0.23 astronomical units).
- Much less is known about **Kepler-9d**; it orbits in only 1.59 days, at a distance of only 0.027 astronomical units from the host star, and has a radius that is 1.64 times the radius of the Earth, putting it into the super-Earth class.[21] At its distance from Kepler-9, the temperature on the star-side of Kepler-9d is a scorching 2,000 K.

A Kepler team led by Jack Lissauer, of NASA's Ames Research Center, discovered that the star Kepler-11 has at least six planets.[22] With orbital periods ranging from 10 to 46 days (Mercury orbits the Sun in 88 days), this is a very compact planetary system. In 2013, using more than 3 years of Kepler observations,[23] Lissauer determined that the three planets **Kepler-11b**, **Kepler-11c** and **Kepler-11f** are super-Earths, with masses and radii of 2–3 Earths—though their actual radii suggest they are more Uranus-like than Earth-like, while **Kepler-11d** and **Kepler-11e** are sub-Uranus-sized planets with radii of 3–4 Earths and masses of 7–8 Earths. The radius of **Kepler-11g** is a bit greater than 3 Earth radii, but the mass of this planet is still poorly constrained. The combination of these radii and masses reveal that all five of these planets have Jupiter-like and Saturn-like densities and must have significant atmospheres composed of hydrogen and helium gasses surrounding planetary cores made of rock or rock and water ice.

The Kepler-33 system has five planets, whose presence was confirmed in 2012 by Lissauer's team.[24] Kepler-33 itself is about the same age and temperature as the Sun, though it is a bit more massive, a bit larger, and about four times brighter.

[20] Holman, M. J., et al. (2010). Kepler-9: A system of multiple planets transiting a Sun-like star, confirmed by timing variations. *Science, 330*, 51.

[21] Torres, G., et al. (2011). Modeling Kepler transit light curves as false positives: Rejection of blend scenarios for Kepler-9, and validation of Kepler-9d, a super-Earth-size planet in a multiple system. *Astrophysical Journal, 727*, 24.

[22] Lissauer, J. J., et al. (2011). A closely packed system of low-mass, low-density planets transiting Kepler 11. *Nature, 470*, 53.

[23] Lissauer, J. J., et al. (2013). All six planets known to orbit Kepler-11 have low densities. *Astrophysical Journal, 770*, 131

[24] Lissauer, J. J., et al. (2012). Almost all of Kepler's multiple-planet candidates are planets. *Astrophysical Journal, 750*, 112.

- Working outwards from the planet with the smallest orbit, **Kepler-33b** is a super-Earth in a 5.7-day orbit with an orbital radius of 0.068 astronomical units and is 1.74 times bigger in radius than the Earth;
- **Kepler-33c** is a Uranus-like planet that orbits in 13.2 days at a distance of 0.12 astronomical units and is 3.2 times larger in radius than the Earth;
- **Kepler-33d** orbits in 21.8 days at a distance of 0.17 astronomical units and is 5.3 times bigger than the Earth, making it a super-Neptune or a Saturn-like planet;
- **Kepler-33e** orbits in 31.8 days at a distance of 0.21 astronomical units and, with a radius that is four times the size of the Earth, is a Neptune-like planet; and
- **Kepler-33f** orbits in 41 days at a distance of 0.25 astronomical units and, with a radius 4.5 times bigger than the Earth, is also a super-Neptune.

Altogether, these five planets are sandwiched closer to Kepler-33 than Mercury is to the Sun, but they are not packed quite as closely together as are the planets in the Kepler-11 system. While none of these planets' masses have yet been measured, Lissauer has shown that for planets in our solar system, the masses of the planets are proportional to the squares of their radii; according to this relationship, a planet whose radius is two Earth radii would have a mass four times greater than that of the Earth. If this relationship applies to the Kepler-33 planetary system, then these planets' masses would range from perhaps three Earth masses for Kepler-33b to 28 Earth masses for Kepler-33d.

In late 2011, William Cochran, of the University of Texas, and his group of Kepler scientists discovered another three-planet system, in orbit around Kepler-18.[25] The planets whiz around the host star in 3.5, 7.6 and 14.9 days. The innermost planet **Kepler-18b** is a super-Earth (2 Earth radii; 6.9 Earth masses) while the outer two planets, **Kepler-18c** and **Kepler-18d**, are Neptune-like planets (17.3 and 16.4 Earth masses, respectively).

ROCKY PLANETS

In early 2011, Kepler mission team scientists announced their first discovery of a rocky planet. This planet was found in the multi-planet system around Kepler-10, which is located about 170 parsecs from the Sun. At the time of the discovery announcement, **Kepler-10b** was the smallest exoplanet discovered via the transit technique and was eye-opening proof that the Kepler telescope was capable of discovering Earth-size planets, though Kepler-10b itself is better described as a super-Earth rather than as an Earth-like planet.[26] From the Kepler data, Natalie Batalha, of San Jose State University, and her team were able to determine that Kepler-10b has a radius 40 % greater than that of the Earth (1.4 Earth radii) and an orbital period of just over 20 h. At a distance of only 0.017 astronomical units from the central star, Kepler-10b has a surface temperature of more than 1,800 K. Using additional measurements taken on the Keck I 10-m telescope on Mauna Kea, Hawaii, the team determined that Kepler-10b has a mass about 4.5 times greater than

---

[25] Cochran, W. D., et al. (2011). Kepler-18b, c, and d: A system of three planets confirmed by transit timing variations, light curve validation, warm-Spitzer photometry, and radial velocity measurements. *Astrophysical Journal Supplement Series, 197,* 7.

[26] Batalha, N. M., et al. (2011). Kepler's first rocky planet: Kepler-10b. *Astrophysical Journal, 729,* 27.

the Earth's mass and a density of about 8.8 g/cm^3, which suggests that the planet is made mostly of rock, with a significant fraction of iron and very little, if any, water. Note, however, that this measured mass is 50 % higher than one would estimate from Lissauer's scaling relationship, suggesting that masses estimated with Lissauer's method may significantly underestimate actual planetary masses. Additional observations of Kepler-10b reveal that it has an extremely reflective surface or atmosphere, comparable to the reflected brightnesses of the most reflective objects in our solar system, Venus and Saturn's icy moon Enceladus. Note, however, that at such a high temperature, this planet cannot have any water, let alone ice, on its surface or in its atmosphere. A few months later, the Kepler team announced that Kepler-10 has a second planet, **Kepler-10c**.[27] Kepler-10c, found in a 45-day orbit, is also a hot super-Earth, with a radius just over twice that of the Earth and temperature of nearly 500 K.

The multi-planet system Kepler-20—a Sun-like star with 91 % of the mass of the Sun and a radius that is 94 % of that of the Sun—includes five planets, three of them super-Earths in size and likely Uranus-like in mass and two of them more Earth-like.

- A team lead by Thomas Gautier III, of the Jet Propulsion Laboratory and Cal Tech, confirmed the existence of three of the planets in late 2011.[28] These planets range in size from 1.91 Earth radii for **Kepler-20b** to 3.07 and 2.75 earth radii for **Kepler-20c** and **Kepler-20d** and have orbital periods of 3.70, 10.85, and 77.61 days, respectively. Gauttier's team used radial velocity measurements to calculate masses for Kepler-20b and Kepler-20c of 8.7 and 16.1 Earth masses and a mass no greater than 20 Earth masses for Kepler-20d. These data reveal that despite the super-Earth size of Kepler-20b, in composition it is most likely a Uranus-like planet.
- A team led by Francois Fressin, of the Harvard-Smithsonian Center for Astrophysics, confirmed the existence of the two Earth-like planets in 6.1 and 19.6-day orbits around Kepler-20.[29] Fressin reports that the radius of **Kepler-20e** is 0.87 Earth radii and the radius of **Kepler-20f** is 1.03 Earth radii, making these two planets nearly identical in size, respectively, to Venus (0.95 Earth radii) and to the Earth. Given their small sizes and proximity to the star Kepler-20, these planets are almost certainly composed, like the Earth and Venus, of rock and iron, are extremely hot (1,000 and 700 K), and are best characterized as hot Earths. Fressin speculates that Kepler-20f could have a thick, water-vapor-dominated atmosphere.

In August 2012, a team led by Joshua Carter, of the Harvard-Smithsonian Center for Astrophysics, announced the discovery of both a super-Earth-like planet and a Neptune-like planet in orbit around Kepler-36.[30]

---

[27] Fressin, F., et al. (2011). Kepler-10c: A 2.2 Earth radius transiting planet in a multiple system. *Astrophysical Journal Supplement Series, 197,* 5.

[28] Gautier III, T. N., et al. (2012). Kepler-20: A Sun-like star with three sub-Neptune exoplanets and two Earth-size candidates. *Astrophysical Journal, 749,* 15.

[29] Fressin, F., et al. (2012). Two Earth-size planets orbiting Kepler-20. *Nature, 482,* 195.

[30] Carter, J. A., et al. (2012). Kepler-36: A pair of planets with neighboring orbits and dissimilar densities. *Science, 337,* 556.

- **Kepler-36b** is the super-Earth, with a radius about 50 % greater than the Earth's radius and a mass 4.45 times greater than that of the Earth, giving it an average density of 7.46 g/cm^3, which means that it is composed of about 70 % rock and 30 % iron. With an orbital period of 13.8 days, Kepler-36b has a surface temperature of about 980 K.
- **Kepler-36c** is the bigger of the two planets, with a radius 3.7 times greater than that of the Earth, a mass equivalent to 8 Earths and a very low, Saturn-like density of 0.89 g/cm^3, making Kepler-36c less dense than water and suggesting that it is composed mostly of hydrogen and helium. Kepler-36c has a 16.2-day orbit, just barely longer than the length of the orbit of Kepler-36b.

The orbits of these two planets place them at distances of only 0.115 and 0.128 astronomical units from the central star. At closest approach to each other, these two planets are only two million kilometers apart, or about five times the distance from the Earth to the Moon. The Kepler-36 planets are a puzzle: How can two planets be made of such different materials and yet exist so close to each other?

Kepler-68 is another multi-planet system with one Earth-size planet, though Ronald Gilliland, of The Pennsylvania State University, and his team do not think that Kepler-68c is Earth-like.[31]

- With an estimated radius of 0.95 Earth radii, the planet **Kepler-68c** is an almost perfect match in size to Venus. Because Kepler-68, however, is almost identical to the Sun and because Kepler-68c orbits in only 9.6 days, this planet is extremely hot and cannot be Earth-like other than in size and mass. Kepler-68 is host to two other known planets.
- **Kepler-68b**, which orbits in only 5.4 days, is a Uranus-like planet with a mass 8.3 times greater than the mass of the Earth. Given its size (2.3 Earth radii), it likely has a core composed partly of rocky materials (like the Earth) and an atmosphere composed of light gases (water, hydrogen and helium).
- The third planet, **Kepler-68d**, is Jupiter-like, with a mass greater than 0.95 Jupiter masses, and has a much larger orbit (580 days), placing it at distance comparable to a location between the Earth and Mars. If Kepler-68d has one or more large moons, those moons could be Earth-size, Earth-like objects and could be in the habitable zone around Kepler-68.

In size, another near-match to the Earth is **Kepler-78b**, whose discovery was announced by Roberto Sanchis-Ojeda, of the Massachusetts Institute of Technology, and his collaborators in September 2013. The Kepler data reveal that Kepler-78b orbits the fairly young (625 million years old) star Kepler-78 in only 8.5 h and has a radius just 16 % greater than that of the Earth. Within a month, two teams had independently made radial velocity measurements from ground-based telescopes and had employed those results to determine the mass and density of Kepler-78b. Francesco Pepe, of the Observatoire Astronomique de l'Université de Genève, and his collaborators[32] used the HARPS-N spectrograph on the 3.57-m Telecopio Nazionale Galileo on the Spanish island of La Palma, while Andrew

---

[31] Gilliland, R. L., et al. (2013). Kepler-68: Three planets, one with a density between that of earth and ice giants. *Astrophysical Journal, 766*, 40.

[32] Pepe, F., et al. (2013). An Earth-size planet with an Earth-like density. *Nature, 503*, 377.

Howard, of the University of Hawaii, and his collaborators used the 10-m Keck I telescope in Hawaii. From the orbital velocity of the star Kepler-78, the two teams were able to calculate the mass of the planet Kepler-78b, which they found is only 69–86 % greater than the mass of the Earth, giving Kepler-78b a density of 5.3–5.6 g/cm^3, which is almost identical to the density of the Earth, leading both teams to suggest that the planet is composed mostly of rock and iron. Because of the extreme proximity to the host star, the surface temperature of Kepler-78b would be 1,500–3,000 K, hot enough to melt the rocks on the surface and making it unlikely that this planet has any atmosphere.

The star identified as KOI 961 (KOI = Kepler Object of Interest; now also identified as Kepler-42) is orbited by three of the smallest exoplanets yet discovered.[33] KOI 961 itself is relatively nearby (39 parsecs from the Sun) and small (14 % the mass of the Sun), while the planets are all smaller than the Earth. Based on Kepler data combined with data from several ground-based telescopes, CalTech's Philip Muirhead and his team determined that **Kepler-42b** has a radius of 78 % of that of the Earth; **Kepler-42c** is a bit smaller, at 0.73 Earth radii, while **Kepler-42d** is the smallest, with a radius of 0.57 Earth radii, making it almost the same size as Mars (0.53 Earth radii). These three planets also have extremely tiny orbits, with Kepler-42d having the longest, biggest orbit, orbiting the central star in 1.9 days at a distance of barely one-sixtieth of the Earth-Sun distance (0.0154 astronomical units). Kepler-42b orbits a bit more quickly, rushing around its orbit in 1.2 days at a distance of 0.0116 astronomical units while Kepler-42c orbits in the least amount of time, completing one orbit in just under 11 h at a distance of 0.006 astronomical units. At these close distances in orbit around this small, faint star, these planets have surface temperatures ranging from 450 to 700 K. Although the planets' masses have not yet been measured, they are almost certainly all less than a few Earth masses, making them all Earth-like planets except that all of them are too hot to have liquid water on their surfaces.

The single smallest known exoplanet is **Kepler-37b**. It and its companions **Kepler-37c** and **Kepler-37d** orbit the Sun-like star Kepler-37 in 13.4, 21.3 and 39.8 days. The radius of tiny Kepler-37b is only 0.3 Earth radii (1,900 km), which is smaller than the radius of Mercury (2,439 km) and just barely bigger than the radius of the Moon (1,738 km). Kepler-37c (0.74 Earth radii) is Earth-size and Kepler-37d (1.99 Earth radii) is a super-Earth. The masses of these three planets are unknown, but the masses of Kepler-37b and Kepler-37c must be comparable to the mass of the Earth or much smaller.[34]

MULTI-PLANET SYSTEMS NEAR MEAN-MOTION RESONANCES

Daniel Fabrycky, of the University of California, Santa Cruz, Eric Ford, of the University of Florida, and Jason Steffen, of the Fermilab Center for Particle Physics in Illinois, together with their other team members, discovered several multiple-planet systems in

---

[33] Muirhead, P. S., et al. (2012). Characterizing the cool KOIs. III. KOI 961: A small star with large proper motion and three small planets. *Astrophysical Journal, 747*, 144.

[34] Barclay, T., et al. (2013). A sub-Mercury-sized exoplanet. *Nature, 494*, 452.

2012 that have special orbital configurations that generate transit timing variations.[35] In these systems, the sizes of the planets' orbits are such that the ratios of their orbital periods are almost exactly ratios of small integers. This means, for example, if one planet orbits in 10 days and another in 20 days, the faster-moving planet would complete two orbits for every one orbit of the slower-moving planet. The planets in these systems have 'commensurable' orbits—the planets in this particular configuration would be in a 2:1 commensurability. Orbital dynamicists refer to the average orbital speeds of planets (or moons) as their 'mean motions' and to such systems as having 'mean motion resonances.' Mean motion resonances are common among moon systems in our solar system and are the result of very strong and sustained gravitational interactions between the moons associated with the resonance. The most famous of these commensurability systems in our solar system involves the inner three large moons of Jupiter. Io, the closest of Jupiter's four big moons to the planet, completes four orbits for every two orbits of a second of Jupiter's big moons, Europa, and every one orbit of the largest moon in the solar system, Ganymede (a 4:2:1 commensurability). Similarly, Saturn's large moons Mimas and Tethys are locked in a 2:1 commensurability.

The planetary systems around Kepler-25, Kepler-27, Kepler-30, Kepler-31, Kepler-48, Kepler-51, Kepler-52, Kepler-53, Kepler-56, and Kepler-57 all have pairs of planets locked into commensurable orbits, with commensurability ratios of 2:1. That is, in all of these systems the inner planet completes two orbits for every one orbit of the outer planet.

- **Kepler-25b** (2.6 Earth radii) and **Kepler-25c** (4.5 Earth radii) orbit in 6.238 and 12.72 days (two times 6.238 is 12.48; the ratio of 12.72 to 12.48 is 1.019; the nearness of this ratio to exactly 1.0 establishes these orbits as commensurable).

- **Kepler-27b** (4.0 Earth radii) and **Kepler-27c** (4.9 Earth radii) orbit in 15.337 and 31.331 days.

- **Kepler-30b** (3.7 Earth radii) and **Kepler-30c** (14.4 Earth radii) orbit in 29.221 and 60.327 days; a third planet in the Kepler-30 system, **Kepler-30d** (10.7 Earth radii), orbits in 143 days.

- **Kepler-31b** (4.3 Earth radii) and **Kepler-31c** (4.2 Earth radii) orbit in 20.856 and 42.638 days. The Kepler-31 system may have two other planets that appear to orbit in 9 and 88 days. If someday confirmed, this system would be even more complicated than the Jovian moon system, having four planets orbiting together with orbital period ratios of 8:4:2:1.

- **Kepler-48b** (2.1 Earth radii) orbits in 4.778 days and **Kepler-48c** (3.1 Earth radii) orbits in 9.674 days. A third planet may exist around Kepler-48, in a 42.896 day orbit,

[35] Fabrycky, D. C., et al. (2012). Transit timing observations from Kepler: IV. Confirmation of four multiple-planet systems by simple physical models. *Astrophysical Journal, 750*, 114; Steffen, J. H., et al. (2012). Transit timing observations from Kepler—III. Confirmation of four multiple planet systems by a Fourier-domain study of anticorrelated transit timing variations. *Monthly Notices of the Royal Astronomical Society, 421*, 2342; Ford, E. B., et al. (2012). Transit timing observations from Kepler. II. Confirmation of two multiplanet systems via a non-parametric correlation analysis. *Astrophysical Journal, 750*, 113; Steffen J. H., et al. (2013). Transit timing observations from Kepler—VII. Confirmation of 27 planets in 13 multiple systems via transit timing variations and orbital stability. *Monthly Notices of the Royal Astronomical Society, 428*, 1077.

in which case this system may have three planets locked into a 18:9:2 commensurability.

- **Kepler-51b** (7.0 Earth radii) orbits in 45.156 days and **Kepler-51c** (5.7 Earth radii) orbits in 85.312 days. A not-yet-confirmed planet orbits in 130.183 days.
- **Kepler-52b** (2.1 Earth radii) orbits in 7.877 days and **Kepler-52c** (1.8 Earth radii) orbits in 16.385 days. A third planet, whose existence has not yet been confirmed, has a radius of 1.8 Earth radii and orbits in 36.446 days.
- **Kepler-53b** (2.9 Earth radii) orbits in 18.649 days and **Kepler-53c** (3.2 Earth radii) orbits in 38.558 days. In this system, a not-yet-confirmed planet orbits in only 9.751 days.
- **Kepler-56b** (3.8 Earth radii) orbits in 10.503 days and **Kepler-56c** (7.9 Earth radii) orbits in 21.405 days
- **Kepler-57b** (2.2 Earth radii) orbits in 5.729 days and **Kepler-57c** (1.6 Earth radii) orbits in 11.609 days.

The smallest of these planets is the super-Earth Kepler-57c. All the others are likely Uranus-like or Neptune-like planets. Notably, very few of these exoplanets in commensurable orbits are as large or larger than Saturn and Jupiter.

Kepler-23, Kepler-24, Kepler-26, Kepler-28, Kepler-32, Kepler-49, Kepler-54, Kepler-55, Kepler-58, and Kepler-59 all are orbited by a pair of planets, and in each system the inner planet completes three orbits in almost the same amount of time needed for the outer planet to complete two orbits (a 3:2 commensurability).

- **Kepler-23b** (1.9 Earth radii) and **Kepler-23c** (3.2 Earth radii) orbit in 7.108 and 10.74 days (three times 7.108 is 21.32; two times 10.74 is 21.48; the ratio of 21.48 to 21.32 is 1.008). A third planet, not yet confirmed, also may exist in this system.
- **Kepler-24b** (2.4 Earth radii) and **Kepler-24c** (2.8 Earth radii) orbit in 8.164 and 12.315 days. Two other planets, not yet confirmed, also may exist in this system.
- **Kepler-26b** (3.6 Earth radii) and **Kepler-26c** (3.6 Earth radii) orbit in 12.28 and 17.25 days. One other planet, not yet confirmed, also may exist in this system.
- **Kepler-28b** (3.6 Earth radii) and **Kepler-28c** (3.4 Earth radii) orbit in 5.912 and 8.986 days.
- **Kepler-32b** (4.1 Earth radii) and **Kepler-32c** (3.7 Earth radii) orbit in 5.901 and 8.753 days. Three other planets, not yet confirmed, also may exist in this system.
- **Kepler-49b** (2.7 Earth radii) orbits in 7.203 days and **Kepler-49c** (2.5 Earth radii) in 10.912 days. Two unconfirmed planets have been found in this system, one orbiting in 2.576 days and the other in 18.956 days.
- **Kepler-54b** (2.1 Earth radii) orbits in 8.011 days while **Kepler-54c** (1.2 Earth radii) orbits in 12.071 days. A third planet may orbit in 20.995 days.
- **Kepler-55b** (2.4 Earth radii) orbits in 27.948 days while **Kepler-55c** (2.2 Earth radii) orbits in 42.152 days. The Kepler-55 system may have three more planets, with orbits of 2.211, 4.617 and 10.199 days.
- **Kepler-58b** (2.8 Earth radii) orbits in 10.218 days while **Kepler-58c** (2.9 Earth radii) orbits in 15.574 days.
- **Kepler-59b** (1.1 Earth radii) orbits in 11.868 days while **Kepler-59c** (2.0 Earth radii) orbits in 17.980 days.

This group of planets includes several Earth-size objects—Kepler-54c and Kepler-59b, a small number of super-Earths—Kepler-23b and Kepler-59c, and a large number of Uranus-like and Neptune-like planets. None of these planets are even half as large, in radius, as Saturn or Jupiter.

Gravitational interactions between orbiting planets can trap planets (or moon systems, or asteroid systems, or moon and planetary ring systems, or planet and asteroid belt systems) into commensurability orbits of many different types, and already the Kepler stars are falling into a broad range of possible configurations:

- In the case of **Kepler-29b** (3.6 Earth radii; 10.336 day orbit) and **Kepler-29c** (2.9 Earth radii; 13.293 day orbit), gravitational interactions have locked the planets into a much-less common 9:7 commensurability (nine times 10.336 days is 93.024; seven times 13.293 days is 93.051);
- For **Kepler-50b** (1.71 Earth radii; 7.812 days) and **Kepler-50c** (2.17 Earth radii; 9.376 days), the commensurability ratio is 6:5;[36]
- The Kepler-60 system is even more unusual. **Kepler-60b** (2.3 Earth radii) orbits in 7.132 days, **Kepler-60c** (2.5 Earth radii) orbits in 8.919 days and **Kepler-60d** (2.6 Earth radii) orbits in 11.901 days. Kepler-60b and Kepler-60c are trapped in a 5:4 commensurability (five times 7.131 equals 35.656; four times 8.919 equals 35.676) while Kepler-60c and Kepler-60d are locked in a 4:3 commensurability (three times 11.901 equals 35.703). Together, the three planets are locked into a 20:15:12 resonance, almost identically to the three-body commensurability involving three of Jupiter's large moons.

Kepler-50b is a super-Earth while all the others of these planets are in the Uranus/Neptune class.

In December 2011, Stéphane Charpinet, of the Université de Toulouse, and his collaborators reported on their discovery, via a very unusual timing variation method, of two super-hot Earths in orbit around a star known as a hot B subdwarf.[37] This star, originally identified as KIC 05807616 and now known as Kepler-70, is hot (27,730 K), tiny (20 % the radius of the Sun) and near the end of its lifetime. Kepler-70 already completed the longest stage of life for a star as a main-sequence star and long ago passed through a stage of stellar old-age known as the red-giant phase. Now, after shedding most of its mass, only the core of the former star remains, and that core pulsates on timescales of hours. The pulsations produce brightness variations in the luminosity output of the star, and those tiny variations were easily measured by Kepler. Charpinet discovered two pulsation modes that are caused by planets very close to the star, both of which are reflecting some of the star's light. The amount of starlight reflected from each planet changes with time and depends on where the planets are in their orbits. During some parts of their orbits, they reflect starlight toward the Earth, while during other phases of their orbits they reflect light away

---

[36]Chaplin, W. J., et al. (2013). Asteroseismic determination of obliquities of the exoplanet systems Kepler-50 and Kepler-65. *Astrophysical Journal, 766*, 101.

[37]Charpinet, S., et al. (2011). A compact system of small planets around a former red-giant star. *Nature, 480*, 496.

from the Earth. Altogether, the dimming and brightening of Kepler-70 because of the reflections from its planets produces measurable brightness variations in the Kepler data.

The two planets **Kepler-70b** and **Kepler-70c** are tiny, high density, hot Earth-size planets that just skim the surface of the star as they orbit. Kepler-70b orbits in 5.76 h at distance of 0.006 astronomical units. This size orbit places it only 29 % of the radius of the star above the star's surface. The day-side temperature of Kepler-70b is above 9,000 K while the night-side temperature is a mere 1,800 K. In size, Kepler-70b most probably is a bit smaller than Venus (0.76 Earth radii) and has an estimated mass just less than half that of the Earth (0.44 Earth masses). Kepler-70c, whose orbital period is 8.23 h, completes 10 orbits for every 7 orbits of Kepler-70b (a 10:7 commensurability). Though a bit further above the surface of the star (the radius of the orbit is 1.64 stellar radii), Kepler-70b is also stunningly hot on both its day side (8,100 K) and night side (1,600 K). Like Kepler-70b, Kepler-70c is also sub-Earth-size in radius (0.87 Earth radii) and mass (0.66 Earth masses).

Charpinet suggests that these two planets were once giant, Jupiter-like planets that orbited at much greater distances from the star. When Kepler-70 swelled up as a red giant, the planets were engulfed by the enormous, swollen atmosphere of the star. The giant planets survived the red-giant phase of the star and may have even triggered the star to shed its outer atmosphere. "During that episode," Charpinet hypothesizes, "the planets may have been stripped down, losing their gaseous layers and being left only with their inner rocky/iron cores."

In 2012 David Nesvorny, of the Southwest Research Institute in Boulder, Colorado, and his colleagues identified a periodic variation in the transit period of the planet **Kepler-46b**.[38] Kepler-46b was discovered by the Kepler mission team as a transiting planet with an orbital period of 33.6 days. Kepler-46c, which orbits Kepler-46 in 57.0 days (putting these two stars very nearly in a 5:3 commensurability) and has a mass of about 0.37 Jupiters, was identified by the Nesvorny team because of the periodic changes its gravity imposes on the orbit of Kepler-46b. A third planet, Kepler-46d, a 1.7-Earth radius super-Earth with a 6.8-day orbital period (in a 5:1 commensurability with Kepler-46b) may also exist in this planetary system.

CIRCUMBINARY PLANETS

Can a planet orbit two stars at once? Yes, provided the orbit of the planet is large when compared to the separation of the two stars; however, if the orbit of the planet is not large enough, the orbit of the planet will be unstable and the gravitational tug-of-war between the two stars and the planet will eject that planet from its orbit. As of the end of 2013, 71 planets (though a few might be larger-than-planet brown dwarfs) had been identified in a total of 53 binary star systems.[39] Almost all of these planets are in wide binary systems:

---

[38] Nesvory, D., et al. (2012). The detection and characterization of a nontransiting planet by timing variations. *Science, 336*, 1133.

[39] Schwartz, R. Binary catalogue of exoplanets. Retrieved from www.univie.ac.at/adg/schwarz/multiple. html. Accessed 12 Nov 2013.

each planet orbits a single star in a small orbit while that star/planet system in turn is in a very large orbit around a second star. Planets orbiting two stars at once, so called circumbinary planets, remain rare, though Kepler has broken the ice and new discoveries are being announced with regularity.

The first evidence for a circumbinary planet in Kepler data was found in 2011 by a team led by Laurance Doyle, of the SETI Institute in Mountain View, California. Doyle used Kepler data to discover that the Saturn-like planet **Kepler-16b** is in orbit around the binary star system Kepler-16AB.[40] The two stars orbit each other in only 41 days and are separated by 0.22 astronomical units (a bit more than half the distance from the Sun to Mercury), while the planet, whose mass is about one-third of the mass of Jupiter, orbits in 229 days at a distance of 0.70 astronomical units (about the distance from the Sun to Venus).

In 2012, two more Saturn-size circumbinary planets, **Kepler-34b** and **Kepler-35b** were found in the analysis of 671 days of Kepler data by a team led by William Welsh, of San Diego State University.[41] Kepler-34b is in a 289 day orbit around a pair of Sun-like stars that orbit each other in 28 days, while Kepler-35b is in a 131 day orbit around a pair of Sun-like stars that orbit each other in 21 days

Kepler-38b orbits a binary system that includes a Sun-like star and a second star that has only one-quarter the mass of the Sun.[42] The two stars orbit each other in 18.8 days. Jerome Orosz and his team found that Kepler-38b orbits the binary star system with an orbital period of 106 days. Kepler-38b has a radius 10 % bigger than that of Neptune (4 Earth radii) and a mass no greater than about the mass of Saturn (about 100 Earth masses).

A team led by V. B. Kostov, of Johns Hopkins University and the Space Telescope Science Institute, used both Kepler data and additional data from the Apache Point Observatory in New Mexico and the Haute-Provence Observatory in France, to discover a gas giant planet around Kepler-64.[43] The binary stars are a mismatched pair; one is about 40 % more massive than the Sun while the other is only about 40 % the mass of the Sun. They orbit each other in 20 days. The large planet **Kepler-64b** (radius 6.2 Earth radii; mass less than 169 Jupiter masses) has a much longer (139 days) and larger (0.64 astronomical units) orbit around the pair of stars than the two stars have around each other. Making this system even more bizarre, a second binary star system, separated by 1,000 astronomical units from Kepler-64, orbits the primary binary star system, putting Kepler-64b inside a quadruple star system.[44]

[40] Doyle, L. R., et al. (2011). Kepler-16: A transiting circumbinary planet. *Science, 333*, 1602.

[41] Welsh, W. F., et al. (2012). Transiting circumbinary planets Kepler-34b and Kepler-35b. *Nature, 481*, 475.

[42] Orosz, J. A., et al. (2012). The Neptune-sized circumbinary planet Kepler-38b. *Astrophysical Journal, 758*, 87.

[43] Kostov, V. B., et al. (2013). A gas giant circumbinary planet transiting the F star primary of the eclipsing binary star KIC 4862625 and the independent discovery and characterization of the two transiting planets in the Kepler-47 system. *Astrophysical Journal, 770*, 52.

[44] Schwamb, M. E., et al. (2013). Planet hunters: A transiting circumbinary planet in a quadruple star system. *Astrophysical Journal, 768*, 127.

HABITABLE ZONE PLANETS

As of the end of 2013, Kepler scientists had discovered a small handful of planets in the habitable zones of their stars. These planets include **Kepler 22b**, **Kepler-61b**, **Kepler-62e**, **Kepler-62f**, **Kepler-69c** and **Kepler-47c**. These planets are described in more detail in Chap. 4.

STATISTICS

Hot planets, these being planets that orbit in less than 50 days (at distances of less than 0.25 astronomical units from their host stars) around stars like the Sun, are the easiest planets to discover. In data obtained with the Kepler telescope, hot planets block more starlight during transits than do more distant planets. In addition, because they orbit quickly they produce transits more often than do slower-orbiting, more distant planets. In radial velocity surveys, hot planets tug harder on their host stars and generate larger and thereby more easily measured radial velocities than planets in larger orbits.

Despite the ease with which they have been found, hot planets, whether large ones like Jupiter or small ones like the Earth, are not, however, found in orbit around most stars. The Kepler data, which specifically includes data almost exclusively for stars fairly similar in size and mass to the Sun, indicate that the bigger, Jupiter-like hot planets are much rarer than smaller planets in short-period orbits. For every one hot Jupiter, Kepler has found evidence for about 100 hot, smaller Uranus-like (two Earth-radii) planets. Andrew Howard, of the University of California at Berkeley, and his collaborators find that the occurrence of all planets bigger than two Earth radii and in short-period orbits is about 17 planets out of every 100 stars. The smaller, Uranus-like planets account for 13 of the 17 planets; the Saturn-like and Jupiter-like planets account for only 4 of the 17.[45] These Kepler results are consistent with results obtained independently in ground-based, radial velocity survey studies that are sensitive mostly to hot Jupiters, though Kepler is finding a few more planets, per hundred stars, than have been found in the other studies. The radial velocity studies find that hot Jupiters appear to be present around about half of 1 % of the stars studied by Kepler and around about 1 % of the stars studied in the ground-based radial velocity surveys.[46]

Howard also finds that the likelihood that a star has one or more planets increases as the mass of the star decreases. That is, little stars are significantly more likely to have planets than are big stars. The numbers reveal that stars that are 60 % less massive than the mass of the Sun are 7 times more likely to have planets than stars that are 50 % more massive than the Sun. In the Milky Way, low-mass stars are far more abundant than more massive stars, and the Kepler survey specifically excluded most very-low-mass stars (below 0.4 solar masses); thus, future surveys that include the least massive stars likely will find at

---

[45] Howard, A. W., et al. (2012). Planet occurrence within 0.25 AU of solar-type stars from Kepler. *Astrophysical Journal Supplement Series*, *201*, 15.

[46] Wright, J. T., et al. (2012). The frequency of hot Jupiters orbiting nearby solar-type stars. *Astrophysical Journal*, *753*, 160.

least several dozen large planets in short-period orbits per hundred stars, rather than only one or two such planets. Furthermore, since the smaller super-Earths and Earth-size planets in short-period orbits are not even part of this statistical study, this study strongly implies that a complete survey that is capable of detecting all of the smaller planets in short-period orbits will find a minimum of several dozen planets in short period orbits per 100 stars. Including the smaller, planets in short-period orbits as well as planets of all types on longer-period orbits, these results seem to suggest that we would find as many or more planets per 100 stars as the number of stars.

Name Index

D.A. Weintraub, *Religions and Extraterrestrial Life: How Will We Deal With It?*,
Springer Praxis Books, DOI 10.1007/978-3-319-05056-0,
© Springer International Publishing Switzerland 2014

Subject Index

A

Angels, 13, 71–72, 84, 93, 98, 113, 117, 143, 148, 150, 156, 157, 162, 166, 167, 171, 180

Articles of faith, 162, 163, 167, 168, 193, 194, 197

Astrometry, 51, 60–62

Astronomical unit, 55, 57, 63, 64, 65

B

Barnard's Star, 43–46, 52, 61

Brahman, 169–174

C

Cosmological principle, 40–41, 79, 130, 209

Crucifixion, 54–90, 94, 108, 120, 129

D

Doctrine of plenitude, 3–5, 17

E

Earth-size planet, 64–66

F

Fermi paradox, 5, 28

G

Geocentric universe, 3

Goldilocks hypothesis, 23

Gravitational lens, 55, 56

Guru, 193–197

H

Habitable zone, 38, 45, 62–68

I

Imaging, 40, 51, 56–58, 68

Incarnation, 84, 85, 89, 90, 93, 95, 97–108, 113, 115, 120, 121, 123–125, 129, 131, 132, 133, 140, 144, 184, 189, 195

K

Karma, 170, 172, 174, 178, 184, 187, 188, 189, 191, 196

Kepler (satellite), 54, 55

L

Lightcurve, 54, 55

M

Microlensing technique, 55–56

O

Original sin, 83–85, 86, 87, 96–98, 104, 105, 107, 108, 116, 124, 129, 130

P

Parsec, 55, 57

51 Pegasus, 43, 50

Pillars of faith, 163, 167

Plenitude, principle of, 12, 15–21, 28, 158, 201

PSR 1257+12, 49, 50, 58, 59

Pulsar planet, 48, 58

D.A. Weintraub, *Religions and Extraterrestrial Life: How Will We Deal With It?*,
Springer Praxis Books, DOI 10.1007/978-3-319-05056-0,
© Springer International Publishing Switzerland 2014

CPSIA information can be obtained at www.ICGtesting.com
Printed in the USA
LVOW09s1736031214

416962LV00007B/220/P